SPEAKING FOR SPOT

Be the Advocate Your Dog Needs to Live a Happy, Healthy, Longer Life

DR. NANCY KAY
Specialist, American College
of Veterinary Internal Medicine

Foreword by John W. Albers, DVM
Executive Director
American Animal Hospital Association

Photographs by Sumner W. Fowler
Illustrations by Beth Preston
Technical Drawings by Alexander Frederick

Trafalgar Square
North Pomfret, Vermont

First published in 2008 by
Trafalgar Square Books
North Pomfret, Vermont 05053

Printed in United States of America

Disclaimer of Liability
This book is not to be used in place of veterinary care and expertise. The author and publisher shall have neither liability nor responsibility to any person or entity with respect to any loss or damage caused or alleged to be caused directly or indirectly by the information contained in this book. While the book is as accurate as the author can make it, there may be errors, omissions, and inaccuracies.

At the time of publication, the Web addresses featured in this book refer to existing Web sites on the Internet. Trafalgar Square Books is not responsible for the content on referred-to sites, including outdated, inaccurate, or incomplete information.

The author has made very effort to obtain a release from all persons appearing in the photographs used in this book and all photographers. In some cases, however, the persons or photographers may not have been known and therefore could not be contacted.

Library of Congress Cataloging-in-Publication Data

Kay, Nancy.
Speaking for spot : be the advocate your dog needs to live a happy, healthy, longer life / Nancy Kay.
p. cm.
Includes index.
ISBN 978-1-57076-405-9
1. Dogs--Health. I. Title.
SF991.K39 2008
636.7'0893--dc22

Statistics included in this book are from: APPMA (pp. 8, 22, 31, 41, 152, 158); AVMA (pp. 28, 47, 168, 176); Pfizer Animal Health/Gallup Organization Dog Owner Survey (p. 12); AAHA (pp. 15, 79, 85, 99); Dogster.com (pp. 133, 227, 229, 232).

Book design by Carrie Fradkin
Jacket design by Heather Mansfield
Typefaces: Novarese, Hypatia Sans, Vag Rounded
10 9 8 7 6 5 4 3 2 1

In memory of my mom, Barbara Silverman.
She would have been so proud.

Contents

Acknowledgments

Naturally, thanks begin with my parents, Barbara and Sheldon Silverman, who tolerated my lifelong infatuation with all creatures great and small and supported my decision to become a veterinarian rather than a "real doctor."

I gratefully acknowledge the veterinary school professors, residents, and interns at Cornell University, who shaped and molded the fledgling that I was some 26 years ago. What a bright, caring, and inspirational group, and how lucky I was to have wound up in their classrooms and clinics. I still remember so many of the things they said (down to the very inflection in their voices).

I am deeply appreciative of my three readers who graciously and thoughtfully invested so much of their time while providing the first edits on this "baby" of mine. My husband Alan Kay, cousin Eve Silverman, and friend Janet Herring Sherman each brought a unique perspective as they pondered my work. They were a huge source of encouragement.

Thanks to my colleagues at my "home away from home," the Animal Care Center in Rohnert Park, California. They graciously offered forth ideas, opinions and support, and patiently answered my many questions pertaining to topics beyond the purview of my area of expertise—internal medicine. They were tolerant of my schedule changes and recent state of "book-induced distraction."

Several truly talented people provided the drawings and pictures that grace the pages of this book. Heartfelt thanks to Beth Preston and Alex Frederick for their fantastic illustrations, and to Tonya Perme, Blair O'Neil, Alan Kay, and Susannah Kay (my daughter who makes me proud as a peacock) for their lovely photographs. Special thanks to Sumner Fowler—a wonderful photographer with an eye for catching those special moments we share with our four-legged friends—for his incredible generosity. I am grateful to the Marin Humane Society for allowing several of their temporary residents to star in this book (including our cover girl, Sandy). And an additional "thanks" to the many human "models" who appear (in some cases, only partially!) alongside their fabulous dogs.

I am profoundly grateful to the "tremendous Trafalgar trio," Caroline Robbins, Rebecca Didier, and Martha Cook. They truly believed in *Speaking for Spot* when so many others were skeptical. Their commitment, expertise, and guidance were nothing short of sensational. They were incredibly patient and gracious with this first-time author, and I feel truly blessed to have been taken under their wing.

Special thanks and love to my husband, Alan, and children, Mickey, Jacob, and Susannah for tolerating this three-year obsession of mine. *Speaking for Spot* chaos and clutter consistently occupied at least half our kitchen table at any given moment. Their invitations and requests were often met with, "I'm working on my book." Yet, throughout this lengthy and arduous book-writing process, they genuinely understood and shared my passion. Alan diligently read every word my computer spit forth and patiently listened to me recite subtitle after subtitle trying to nail down just the right one. He was a constant source of constructive feedback—and was correct most of the time. He tolerated me dragging a laptop everywhere we went and crawling out of bed much too early in the morning to squeeze in some writing time. Most importantly, when the prognosis for publication appeared bleak or I struggled to complete a particularly difficult section, Alan was the consummate gentle, yet effective coach cajoling me to remember why I wanted to write this darned book in the first place.

Finally, I am deeply grateful to the many thousands of people who have given me the chance to provide medical care for their four-legged family members. It has been a privilege (there's no greater profession on earth!) and from them (both clients and patients) I've received a fine education, beyond anything I could have ever learned in the classroom.

Foreword

Without a doubt, the best veterinary care for pets occurs when decisions are made and treatment plans are developed as the result of a strong collaborative effort between the pet's owner and the pet's veterinarian.

Until dogs develop far greater cognitive and reasoning skills, until they greatly expand their communication abilities, and until they can search the Internet for medical information on their own, they will need others to be their advocates when it comes to their veterinary care. Both the owner and the veterinarian share this responsibility.

It sounds easy to say, "Just do what's in my dog's best interests." Making decisions for a pet's care is not rocket science, but on the other hand, it can be complex, confusing, emotional, and intense. Fortunately, Dr. Nancy Kay provides an enormously helpful guide to decision-making, from selecting a pet, to choosing a veterinarian and a practice, to making decisions about care and treatment. Her advice and suggestions will help all pet owners understand the issues and the process involved in medical decision-making.

While we tend to focus on those decisions necessary in cases of serious or life-threatening illness, advocacy is just as important, as Dr. Kay points out, when it comes to routine care. When and how often to vaccinate, and for what diseases, is a significant decision. So are choices about spaying and neutering, parasite control, and dental care. As in the case of serious illness, shared decisions, made after respectful communication between owner and veterinarian, are almost always best.

In previewing *Speaking for Spot* I found myself wishing that Dr. Kay's book had been available when I was in practice, and read by the clients who chose me to be their pet's veterinarian. I have to admit, as well, that I, too, could have benefited from this book—as could some of my colleagues in practice today. If you think your veterinarian might be in that category, you might want to share this book with her. Working together, you and your veterinarian will ensure that your pet gets the very best care possible.

John W. Albers, DVM
Executive Director
American Animal Hospital Association

"A dog doesn't care if you're rich or poor, big or small, young or old. He doesn't care if you're not smart, not popular, not a good joke-teller, not the best athlete, nor the best-looking person. To your dog, you are the greatest, the smartest, the nicest human being who was ever born. You are his friend and protector."

—LOUIS SABIN, AUTHOR OF *ALL ABOUT DOGS AS PETS*

Introduction

If you are spending time with this book, you likely care deeply for your dog. You are not alone in the way you feel—far from it! Today the human-animal bond is stronger than ever. Perhaps, the more tumultuous the world around us, the tighter we cling to our beloved pets. They soothe us with their predictability and unconditional love, and they consistently give in excess of what they receive.

As a veterinarian, I have the good fortune of meeting and interacting with people every day who are experiencing the richness that animals bring into our lives. The human-animal bond is a profound source of inspiration for most veterinarians, myself included. In fact, it is a major source of inspiration for this book.

My book-writing journey was jumpstarted by a breast cancer scare of my very own. Three months of my life were blown off course by a myriad of diagnostic tests and conflicting opinions, all triggered by an abnormality on a routine mammogram. I certainly experienced a good deal of apprehension and uncertainty, but from the time my ordeal began until the moment it ended, the overriding feeling I had was that I was incredibly fortunate because I knew how to be a clever and effective medical advocate for myself. I knew how to ask important questions, find the right doctors, and make informed choices. An unnecessary surgical procedure was avoided because I did more than simply follow one doctor's recommendations. Rather than allow myself to be buffeted about in choppy medical waters, I managed to remain steady at the helm. I was able to transform what could have been an impossible medical maze into a clear pathway.

Thankfully, my medical saga had a happy ending (not a trace of cancer) as well as a substantial silver lining—it planted the seed from which this book germinated. Having benefited profoundly from my own medical advocacy skills, I had a strong desire to teach others how to do the same. Given what I do for a living, it made sense to begin close to home. *Voila*, the birth of *Speaking for Spot*!

My hope is that this book will help you become a highly effective medical advocate for your four-legged best friend. Who knows, you might just become a better medical advocate for yourself in the process.

Note to the Reader

Before you sample the pages of this book, I'd like to clarify some of the terminology you will encounter. Unlike human medical doctors who need to think only in terms of their "patient" (and in school, they need only learn the medical nuances of one, rather than multiple species), we in the veterinary profession are doubly blessed with both "patients" and "clients." The "patient" is the cuddly creature before us, and the "client" is the human at the other end of the leash. As veterinarians, we practice medicine on our "patients," and practice social work (sometimes of a hospice nature), instructional assistance for at-home dog-care techniques, financial planning, and even "psychotherapy" for our "clients." And you wonder why we are always running behind schedule! If someone goes to veterinary school because they love horizontal creatures, but aren't so crazy about the vertical variety, they're in for some major job dissatisfaction.

> QUICK REFERENCE
> **Anatomical "Roadmap"**
>
> If you are like most people, you might be uncertain as to exactly where the kidneys, gall bladder, pancreas, and other organs are located within your dog's body. As you read this book and a particular organ or body structure is mentioned, I encourage you to refer to the drawings located in Appendix II. p. 283. They will serve as your anatomical "roadmap" and help you learn more about how your dog's body functions.

In the veterinary world, the terms "office," "clinic," and "hospital" are used interchangeably, and do not necessarily specify the kind or level of care one might receive. However, for ease of translation from what we are all familiar with in human care to what we may not recognize in veterinary care, I have used the term "office" or "clinic" when referring to the workplace of your family veterinarian (the one you visit on a regular basis) and "hospital" when discussing an emergency or specialty care facility.

For the sake of my writing sanity and your reading sanity, I've made some gender decisions. Throughout the pages of this book, I've chosen to stick with the female gender when referring to veterinarians. Lest my male readers despair, keep in mind that, currently, the vast majority of veterinary school graduates are women. Clients are also female. This leaves the male gender for the dogs in my book—other than in discussions about spaying and pregnancy, of course.

Many of us in the veterinary profession are currently dancing around the term "dog owner." People invariably agree that we don't "own" our human children. Similarly, some take offense at the notion of "owning" the canine members of their family, and others might argue that we don't really own our dogs, because in reality, our dogs own us! I've yet to come up with a replacement term for "owner" that, to my way of thinking, satisfactorily captures the true meaning of who we are in relationship to our dogs. Until I do, I am avoiding the use of "owner" whenever possible and opting instead for "companion" and "caregiver." If more suitable terminology pops into your head as you read this book, I'd love to have you share it with me. (You can contact me through my Web site www.speakingforspot.com.)

Just as with humans, in the world of veterinary medicine one can practice Western medicine, Eastern medicine, or a combination of the two. *Western medicine*, also referred to as "conventional medicine," is commonly practiced in much of the world today (think antibiotics, surgery, blood pressure medication). *Eastern medicine*, also often known as "complementary" or "alternative medicine," encompasses massage therapy, acupuncture, chiropractic, acupressure, herbal therapy, homeopathy, and a host of other mind- and body-healing techniques.

My veterinary school, internship, and residency training focused exclusively on Western medicine. Since then, I've received a thimbleful of training in Eastern philosophy and technique. I certainly believe in its power and have had the good fortune of working collaboratively with some highly adept practitioners. However, my Western medicine background will be readily apparent as you read *Speaking for Spot*. Keep in mind that just as the comprehension of meditation techniques could benefit the practice of *any* religion or philosophy, the medical advocacy techniques you learn here will serve you well regardless of the genre of medicine you choose for your dog.

I sometimes daydream about a fantasy world for dogs where all veterinarians are Dr. Doolittles, capable of understanding "dog-speak" and the wishes and desires of our patients. We'd know how they were responding to our attempts to make them feel better. Just think how easy it would be to come by a diagnosis—the limping dog could direct us to the sore spot; he could tell us why he feels like eating only half his dinner rather than polishing off the whole bowl. Most importantly, we could know with certainty which dogs still have the desire to "fight the good fight," and which ones are wishing for a peaceful and humane end to their suffering. Wouldn't that be an incredible gift!

Unfortunately, only movie stars Rex Harrison and Eddie Murphy are gifted with such Dr. Doolittle talents, so I must settle for something a tad more realistic. In this less whimsical fantasy there is—attached by a leash to *every* dog—a motivated and effective human medical advocate. Motivation is the easy part—most people intrinsically want to provide their dogs with the best medical care. But, becoming an *effective* medical advocate presents a much greater challenge. Coaching and education are typically required. If you are in need of such guidance and instruction, you've come to the right place—*Speaking for Spot* will guide your transformation into a new and improved advocate for your best friend. So turn the page, dear reader. You, your dog, and your veterinarian will benefit from what you are about to learn.

"He is your friend, your partner, your defender, your dog. You are his life, his love, his leader. He will be yours, faithful and true, to the last beat of his heart. You owe it to him to be worthy of such devotion."

—AUTHOR UNKNOWN

1

Medical Advocacy 101

You've now found yourself totally enamored of a hairy, four-legged creature and will want to do the very best possible job you can to care for him. You feed him high quality food, and provide him with regular exercise and plenty of play time. You even let him sleep on the bed! That's the easy part. The hard part comes when you need to make significant medical decisions that will impact your dog's health. You may not have thought much about it at the time, but when you accepted your dog as part of the family you agreed to take care of him both in sickness and in health. You "signed" an unwritten contract, whereby you accepted "power of attorney" to act for your dog and be willing and able to make medical decisions on his behalf.

You "signed" an unwritten contract, whereby you accepted "power of attorney" to act for your dog and be willing and able to make medical decisions on his behalf.

Your role now becomes much more than caretaker and friend. In exchange for that wagging tail and unconditional love, you now become your best friend's medical advocate. Maintaining your dog's health means gathering information, making important choices, dealing with illness, and potentially tackling the question of euthanasia. Welcome to the toughest part of sharing your life with a dog.

Consider the example of Riley, a 12-year-old German Shepherd mix who is cared for and adored by the Johnson family. Riley is the family exercise partner, newspaper retriever, nanny, bedmate, and comic relief in his busy household. He's also in charge of training the newest family member named Bubba, a 10-week-old Shepherd puppy. But recently, the normally ravenous Riley has been leaving some food in his bowl, tiring on his walks, and vomiting. The family veterinarian has determined from some blood tests that Riley has kidney failure. She's recommended that Riley be hospitalized for round-the-clock intravenous fluid therapy as well as an abdominal ultrasound examination, specialized blood tests, and a possible kidney biopsy to determine the cause. Although the outlook is bleak, it certainly isn't hopeless. The cost for all this is estimated at $3,000 to $5,000.

39% of US households own at least one dog.

On first hearing the news, the Johnsons are devastated and confused. They had no idea Riley was so sick. They ask if they had brought him to the clinic sooner, could all of this have been avoided? The three Johnson children have never known life without their beloved pet. How will they respond to this news? Who will model civilized "doggie" behavior for Bubba? The Johnsons aren't sure the recommended care will be affordable. They don't want to throw in the towel too early, nor do they want Riley to suffer. Should they get a second opinion? Is it reasonable to proceed with the recommended diagnostic tests and treatment with a 12-year-old dog?

The veterinarian asks if they have any questions. Questions? The Johnsons don't even know where to start. They want to do what's best for Riley. The problem is they aren't exactly sure what that is. They feel incapacitated by their lack of medical knowledge and their emotional turmoil.

I suspect that some of you are reading this book because you have a "Riley" of your very own and are perhaps experiencing many of the Johnsons' struggles and emotions. If this is the case, now is the perfect time to learn how to effectively "speak" for your dog.

Why Your Dog Needs a Medical Advocate

Gone are the days when you simply followed your vet's orders and asked few, if any questions. The vet is now a member of your dog's health-care team, and you get to be the team captain!

Gone are the days when you simply followed your vet's orders and asked few, if any questions. The vet is now a member of your dog's health-care team, and you get to be the team captain! Your job description has evolved from receiving and following doctor's orders to processing and making decisions. This is no easy task given the volume of information and number of diagnostic and treatment choices available today. Consider the fact that in the human field, medical knowledge doubles approximately every seven years. I suspect that this is true in veterinary medicine as well. Many positive changes in the veterinary profession such as ultrasound, advanced surgical procedures, cancer treatments (the list goes on and on) have created even greater need for people to act as their dog's medical advocate. There are far more choices than ever before.

In addition to the family vet (the veterinary version of our primary care physician), people now have access to a barrage of specialists, including internists, cardiologists, neurologists, dermatologists,

SECRET FOR SUCCESS

Educate Yourself

The best way to become a superb canine medical advocate is to "get smart" about your dog's health. I know, I know...if you were interested in medicine, you would have gone to medical school. I'm not saying you need to become a Dr. Einstein—only that you need to accumulate and understand some basic information. This will allow you to participate in making sound medical decisions. Say your dog has kidney failure. You don't need to understand the microscopic physiology of salt secretion within the kidney tubules (although it is fascinating stuff). However, your pup will benefit immensely if you understand how diet, supplemental fluids, and treatment of the high blood pressure that accompanies kidney failure can impact his medical outcome.

Talk to Your Vet

A great way to begin gathering the knowledge you need is by asking your vet lots of questions (see p. 131). Assuming you've selected your veterinarian wisely (see chapter 3), she will appreciate your interest and involvement. Like most veterinarians, I delight in explaining things to well-informed clients because the more they understand, the greater the likelihood we will accomplish our mutual goal—taking really good care of their dog.

When it comes to expanding your knowledge, don't stop with your veterinarian. Take advantage of the entire veterinary staff. Count on them for dog food recommendations, as well as information about vaccinations, heartworm prevention, and flea and tick control.

Read

Another fabulous way to become a savvy medical advocate is to get your hands on current, reliable medical information, and read, read—then read some more. When you come across material that confirms something you've already heard about, it will feel as though you've received a reassuring second opinion.

When I explain a medical situation to a well-read client, it is as though I can skip elementary school and go directly to junior high—or even high school. The level of understanding that can be achieved during our discussion is invariably greater. There are a number of good books on canine health—see some of my recommendations on p. 376, but ask your veterinarian for hers, as well. I encourage you to keep several (this one included, of course!) on your bookshelf.

Surf the Web

The Internet can be an incredibly useful source of information. I happen to enjoy hearing what my clients are learning online. I sometimes come away with valuable new information, and I'm invariably amused by the extraordinary things they tell me. (Who knew that hip dysplasia is caused by global warming?) Surf to your heart's content, but be forewarned—not all veterinarians feel as I do. Some detest the practice because of all the "whackadoodle" Web sites their clients access. They don't want to spend valuable office time talking clients out of crazy notions and reining them in from wild goose chases.

How can you figure out which sites offer trustworthy information? It's not always easy to tell, but here are some general guidelines:

1. Ask your veterinarian for her Web site recommendations. She might want to refer you to a site that will reinforce or supplement the information she's provided.
2. Veterinary college Web sites invariably provide reliable information. Search for them by entering the words "veterinary college" after the name of the disease or condition you are researching.
3. Web addresses ending in ".org," ".edu," and ".gov" represent nonprofit organizations, educational institutions, and government agencies, respectively. They will likely be objective sources.
4. When your dog has a breed-specific disease, pay a visit to the site hosted by that particular breed association.
5. If you come across business-sponsored Web sites that stand to make money when you believe and act on what they profess (especially if it involves buying something), immediately move on!
6. Be ever-so-wary of anecdotal information. It's perfectly okay to indulge yourself with remarkable tales (how Max's skin disease was miraculously cured by a single session of aromatherapy), but view what you are reading as fiction rather than fact. As fascinating as these *National Enquirer*-type stories may seem, please don't let them significantly influence the choices you make for your dog.
7. Don't forget about online support groups pertaining to particular dog diseases. I love them because they offer a variety of opinions, and sometimes my clients pick up nifty new ideas and techniques that I get to learn about as well. (Be sure to run anything new by your vet before trying it out on your dog.) In addition, online groups can be a source of tremendous emotional support—always a good thing for the humans involved.

ophthalmologists, radiologists, surgeons, nutritionists, and dentists. Other veterinarians specialize in alternative, or complementary, medicine that encompasses acupuncture, chiropractic, homeopathy, and herb therapy. And, certified veterinary rehabilitation specialists can profoundly and positively impact recovery time and comfort level for dogs suffering from arthritis or recovering from back or joint surgery.

More than 75% of owners say their dog's health is as important to them as their own.

Veterinary technology is also keeping pace with its human counterpart (see chapter 5, p. 77, for details). MRI and CT scanners are now options as are 24-hour veterinary critical care facilities. Dialysis is available for the dog with kidney failure. Chemotherapy and radiation therapy have turned many canine cancers from a death sentence into a treatable disease. Advances in veterinary health care are allowing dogs to live longer lives, so it makes sense that veterinarians are recognizing and treating far more age-related disorders such as kidney failure, heart disease, arthritis, and cancer. Vaccinations are available to prevent 13 different canine diseases. Hundreds of prescription diets exist for dogs with diabetes, kidney, liver, gastrointestinal, heart, skin, and joint diseases. It seems there are more Web sites on dog health issues than there are dogs. Whew! No wonder your dog needs a medical advocate!

Emotional attachment just about guarantees difficult, sometimes gut-wrenching medical decisions will need to be made down the line.

What are the chances that you'll *never* be called upon to act for your dog in a medical situation? Probably the same as winning the lottery. I'd love to be more reassuring, but the fact is, after almost three decades as a practicing veterinarian, I know only a handful of dogs who maintained a lifetime of completely good health and vigor right up to the moment of gently and painlessly passing away in their sleep. Sooner or later, almost every dog becomes sick, and for the majority of people, emotional attachment just about guarantees difficult, sometimes gut-wrenching medical decisions will need to be made down the line.

Why the Advocate Needs to Be YOU

Now you know why your dog needs an advocate, but why must it be you? Why not pass this responsibility on to your veterinarian, or your girlfriend who is a nurse, or your first cousin who happens to be a pediatrician? After all, their medical backgrounds seem to give them a clear advantage.

What you may not realize is that *you* are the most qualified of all because absolutely, positively *no one* knows your dog better than you do. You are acutely familiar with the nuances of his daily routine and behavior. You are the one who has the best idea what his soulful expression is meant to convey. Deep down you know better than any veterinarian, medical doctor, nurse, well-intentioned relative, boyfriend, girlfriend, or busybody neighbor what will most likely cause your dog to wag his tail in triumph. You are also the one who knows whether or not your buddy likely wants to "keep fighting the good fight" when the going gets tough.

The person who is willing to step up to the plate when it comes time to make medical-care choices is far more likely to walk away with peace of mind than one who has deferred to others' opinions.

By all means, solicit opinions from experts and people you trust, but, for your dog's sake, be certain that final decisions come from your own mind—and heart. The person who is willing to step up to the plate when it comes time to make medical-care choices is far more

likely to walk away with peace of mind than one who has deferred to others' opinions. And, even when the outcome is poor, the active participant derives comfort from knowing she had nothing but the best of intentions for her beloved dog. She can take solace in knowing she put forth her best effort to make informed decisions. When the decision-making responsibility is relinquished and things don't turn out well, it's very easy to feel you have abandoned your best buddy during his time of greatest need. And, when death is the outcome, it can be extremely difficult—and sometimes impossible—for anyone who "bowed out" to move past his or her grief.

Advocacy Starts Early

Consider the following scenario—one I can assure you every small animal veterinarian has endured. It's time for a puppy's first examination. When the vet enters the room, she encounters an incredibly cute, wiggly, waggly fluff ball and his adoring, newfound humans, who are beaming with pleasure. The vet listens to the pup's heart with a stethoscope and detects a heart murmur, darn it. Maybe she heard incorrectly, so she listens again. Yup, it's there for sure, and it's a loud one. One test leads to another until the birth defect has been clearly defined. It's a heart anomaly that will, with certainty, result in a profoundly shortened lifespan. Smiles turn to tears and heartache.

Before You Fall in Love

Your primary goal is a healthy dog, so doesn't it make sense, if possible, to start out with a dog *you know* is healthy? Don't fool yourself into believing that, like a new appliance, you'll simply be able to return the new dog if problems are discovered. Please don't be tempted by breeders and adoption agencies who guarantee a replacement pup if yours is discovered to have flaws. They'll make good on their promise, but you'll be hard-pressed to relinquish the new love of your life. Trust me when I tell you that it typically takes no more than four minutes and 23 seconds for the average person to fall hopelessly in love with a dog. And four minutes and 22 seconds just isn't enough time to make sure all the necessary medical checkups have been performed—or if they have, to study the results! Do yourself a favor and protect yourself from a broken heart by getting the information you need before—not after—you adopt or purchase a new dog.

58% of dog owners report they visit their pet's veterinarian more than they do their own physician.

Whether you are adopting from a shelter or purchasing a dog from a breeder, make sure that a veterinarian has evaluated your prospective pup, *before* you meet him! Then make the effort to learn what it is the veterinarian discovered. Don't be seduced by the classified ad that says, "vet-checked" because this says nothing about the veterinarian's actual findings. Try to speak directly with the veterinarian who performed the exam, or at least read through the official medical record. You want to be sure beforehand that the little guy now chewing on your shoelace doesn't have a cleft palate, heart murmur, hernia, or any other congenital health issue.

I realize that it's not always possible to have a dog "vetted" in advance of adoption. If you find yourself in this situation, it's best to schedule an exam with a veterinarian as soon as you can—preferably on your way home from picking up your new pup.

Special Considerations for the Purebred Pup

Let me begin by saying that I strongly encourage you to find the next love of your life at a pet shelter or a breed rescue service that finds homes for displaced purebred adult dogs. (In addition, I recommend you work with a shelter or breed rescue organization that performs an extensive behavior evaluation on each dog so you have a better chance of finding just the right match.) I recognize, however, that for some a specific breed fits the bill best, a puppy rather than an adult is desired, and a particular breeder is the source that has been recommended by friends, family, or other dog lovers. If this is the case, there are important details to consider before making a puppy purchase.

Here is a situation that every veterinarian can relate to. A one-year-old Labrador Retriever has just started training for his career as a field trial dog. Just a week into the training program, the pup comes up stiff and sore with pain in both front legs. X-rays show that he has an elbow abnormality commonly inherited in Labs. Surgery will be required, and there is significant potential that he will have lifelong arthritis in both elbow joints. His future as a field trial dog has just unraveled. The client is disappointed because the pup's dam and sire (parents) had both been officially certified and found to be free of elbow disease. A little bit of retrospective research, however, reveals that several of the dog's aunts and uncles had unfavorable elbow screenings.

If you plan to share your life with a purebred dog, before you so much as peek at a puppy, learn as much as you possibly can about potential breed-specific inherited medical issues. The

If you plan to share your life with a purebred dog, before you so much as peek at a puppy, learn as much as you possibly can about potential breed-specific inherited medical issues.

more you know, the more likely you are to choose a puppy free of, and unlikely to develop, such inherited health issues. A word of warning: don't dare rely on the proverbial, "None of my dogs have ever had *that* problem."

A conscientious breeder will offer forth official paperwork rather than verbal reassurances.

A conscientious breeder will offer forth official paperwork rather than verbal reassurances. Study the documents to find out if the parents have been officially and *favorably* screened for the appropriate breed-related diseases. Don't stop there. Take the time to get the same information about the dam's and sire's littermates (all those aunts and uncles). More and more, we are learning the best way to ensure a puppy will be free of inherited diseases is by looking for squeaky-clean health screening results for all his aunts and uncles *in addition* to his parents. What if the dam and sire each had 10 littermates? This means that you are going to be looking at *a lot* of paperwork!

A dog can be officially certified free of specific inherited diseases in a number of ways. First, you need to do some homework to figure out which are the most appropriate screening tests for the breed you are investigating (see below). For example, auscultation of the heart (listening with a stethoscope) may be all that is needed to screen for an inherited heart defect in one breed of dog. In another breed, an ultrasound evaluation of the heart may be the test of choice. How are you to know which screening certification to be satisfied with? Here are some steps to help you figure it out:

1. Research which diseases are common or potentially inherited in the breed you fancy. Potential sources of information include your veterinarian, reputable breeders, the breed association, the American Kennel Club, inherited disease registries, and reference materials found online or in current publications.

2. Find out which screening tests are considered most reliable to check for such diseases and which family members should be screened for them (puppy, dam, sire, aunts, uncles). You can get this information by talking with the experts. Begin with your own veterinarian as well as those who specialize in the health issue of concern. Compare what they have to say with reputable representatives from the breed association (not just the person trying to sell you a puppy).

   Let's take Newfoundlands, for example. This breed is predisposed to subaortic stenosis, an inherited heart defect. Learn which heart-screening test is best—listening to the heart with a stethoscope or performing an echocardiogram—by talking with your veterinarian and a board certified veterinary cardiologist, or if that is difficult, a knowledgeable "Newfie" nerd or two. With other breeds, a bone disease might be of concern, so ask for advice from a board certified veterinary surgeon; for eye disease, a board certified ophthalmologist. See chapter 5 for help finding veterinarians who specialize in different diseases.

3. Learn how to interpret test results. For example, when it comes to hip screening, the Orthopedic Foundation for Animals (OFA) and the University of Pennsylvania Hip Improvement Program (PennHIP) are the bodies that evaluate X-rays to determine the presence and severity of hip disease. You'll want to know what test result or "ranking" is acceptable (this varies from breed to breed). What about the puppy with three of 18 aunts and uncles with "poor" hip ratings? Rely on the same expert you consulted in Step 2 for guidance.

4. Ask the breeder to provide you with all the paperwork (certificates documenting the results of official examinations) you need to eval-

uate in order to do the best possible job landing yourself a healthy puppy.

QUICK REFERENCE
Four "Must Dos" before Purchasing a Purebred Dog

❶ Research the inheritable diseases in the breed (see sidebar, p. 20).

❷ Learn about the preferred disease screening tests.

❸ Find out how to interpret the screening test results.

❹ Ask the breeder to provide documentation of all of the screening test results.

Please allow me one paragraph to get on my soapbox. Lots of "puppy mills" (large scale breeding operations that produce puppies for profit—often inhumanely) stay in business because people purchase purebred pups without doing the research recommended above. So, I beg of you, don't even think about purchasing a puppy online, sight (and site) unseen. And, be extremely cautious about an impulsive pet store purchase, where you still need all the same guarantees I've outlined for a direct purchase from a breeder. By buying online or from a pet shop, you may be inadvertently committing the next 10 to 15 years of your life to taking care of an adorable, but inherently unhealthy, product of a puppy mill. One less purchase from puppy mills, even indirectly, is one step closer to their extinction.

Now that You've Fallen in Love

After you've bought or adopted that adorable little shedding and shredding beast, there's more work to be done with regard to potential breed-related health issues (even if you've already done all of the necessary pre-purchase research). Rather than sit around and wait to see if a disease is going to rear its ugly head, consider taking a proactive approach. When it comes to inherited diseases, early detection and intervention can favorably affect long-term outcome.

QUICK REFERENCE

Official Canine Disease Registries

CERF: Canine Eye Registration Foundation:
www.vmdb.org/cerf.html

GDC: The Institute for Genetic Disease Control in Animals:
www.parrotcreek.com/gdc

OFA: Orthopedic Foundation for Animals:
www.offa.org

PennHIP: The University of Pennsylvania Hip Improvement Program:
www.pennhip.org

VMDB: Veterinary Medical Database:
www.vmdb.org

Let's say you've adopted a young German Shepherd mix. Have his hips X-rayed when he is still a youngster to check for hip dysplasia (a malformation of the ball and socket joint of the hip). Surgery can sometimes be performed at a young age to prevent or lessen the debilitating arthritis that results from this disease. Perhaps you've acquired a Cavalier King Charles Spaniel—one of the most endearing breeds on the face of the earth, but markedly predisposed to an inherited disease affecting the mitral valve of the heart. An ultrasound evaluation of the heart when your dog is still a youngster is an invaluable way to establish baseline parameters. Down the road, you can use them for comparison to help determine any progression of the disease.

These abnormalities are easy to investigate in advance when you have a purebred pup, but up until recently, all you could often know with certainty about your mixed-breed dog was that "Mom" was a young unwed mother, and the father was a bit of a loose cannon. Nowadays, companies offer genetic testing to figure out which types of purebred dogs went into the making of your mutt. So, it's ultimately possible to investigate breed-related health issues in *any* dog.

QUICK REFERENCE

Commonly Inherited Diseases

BLOOD ABNORMALITIES
Autoimmune hemolytic anemia
Hemophilia
Von Willebrand's disease

ENDOCRINE DISEASES
Addison's disease
Cushing's disease
Diabetes mellitus
Hypothyroidism

CANCER
Lymphoma
Malignant histiocytosis
Mast cell cancer
Osteosarcoma

EYE DISEASES
Cataracts
Ectropion
Entropion
Glaucoma
Keratoconjunctivitis sicca (dry eye)
Lens luxation
Retinal dysplasia

GASTROINTESTINAL DISEASES
Colitis
Copper storage disease
Gastric torsion
Hepatitis
Inflammatory bowel disease
Perianal fistula

GENITOURINARY TRACT DISEASES
Bladder stones
Cryptorchidism
Fanconi syndrome

HEART DISEASES
Aortic stenosis
Cardiomyopathy
Degenerative vavular disease
Mitral valve dysplasia
Patent ductus arteriosus
Pulmonic stenosis
Tricuspid valve dysplasia

NEUROLOGICAL DISEASES
Cervical vertebral malformation
(Wobbler syndrome)
Deafness
Degenerative myelopathy
Epilepsy
Hydrocephalus
Myasthenia gravis

ORTHOPEDIC DISEASES
Elbow dysplasia
Hip dysplasia
Intervertebral disk disease
Osteochondrosis
Patellar luxation

SKIN DISEASES
Acne
Allergic dermatitis
Demodecosis
Pyoderma
Sebaceous adenitis
Seborrhea

GRETL, a gorgeous and keenly intelligent Bernese Mountain Dog, passed away when she was only eight years old, her life cut short by cancer. Her death left a huge void in her devoted companion Kathie's life, as well as in the lives of the elementary school children who were reading to Gretl once a week as part of their "Share a Book" program, designed to help them progress in their reading skills. Kathie advised me that, of Gretl's five littermates, three had already succumbed to cancer. And, a fourth had died from an undefined illness.

Imagine my trepidation when I was asked to evaluate Bailey, Gretl's last living littermate. Wouldn't you know it—cancer within the abdomen was discovered. At the time of this writing, we are treating Bailey with chemotherapy and hoping for the best.

Please Don't "Litter"

Any discussion about medical advocacy basics would be incomplete without emphasizing the health benefits of spaying and neutering. Of greatest importance is the prevention of undesired pregnancies. They also prevent ovarian, uterine, testicular, and certain types of prostate disease. When performed before a female dog's first heat cycle, there is the added benefit of greatly decreasing the risk of breast cancer. Finally, neutering may prevent undesired behaviors caused by "testosterone toxicity" (roaming away from home, fighting with other dogs) that can result in traumatic injury. Unless you have a valuable purebred dog that you are hoping to breed, begin talking about spaying or neutering at the very first vet visit. And, if you choose to breed your dog, don't

75% of owned dogs are spayed or neutered.

forget about the potential health benefits of doing so later in life once his or her breeding career is over.

"My basic principle is that you don't make decisions because they are easy; you don't make them because they are cheap; you don't make them because they're popular; you make them because they're right."

—REVEREND THEODORE M. HESBURGH

Medical Decision-Making You and Your Dog Can Live With

Making significant medical decisions can be mighty challenging, even for those with medical experience. Why must it be so difficult? For starters, we all love our dogs dearly, and the thought of making the wrong choice on their behalf is a dreadful one indeed. It would be different if dogs had a voice in it; instead the burden rests 100 percent on our imperfect human shoulders.

With so many factors to consider and so much medical information to comprehend and digest, deciding what to do can be really tough. I'd love to be able to give you a secret recipe, but the main ingredient doesn't exist—namely the crystal ball that would preview the outcome of our choice. In lieu of a magic recipe, here is some sound advice and reassurance to help keep you on track.

How to Make Choices

I've seen clients deal with making medical decisions in a number of ways. First, there is the "eenie-meenie-miney-mo" approach. It's certainly quick and easy, but as tempting as this method sounds, I discourage it at all costs. Making decisions this way is a huge gamble, and your dog's health is what's at stake.

Second, there is the abdication of responsibility; in other words, solicit someone to do it for you. I'm not crazy about this strategy either, for two reasons. No matter how knowledgeable your designated pinch hitter is, there's no way in the world he or she knows your dog well enough and what's ultimately in his best interest. If the end result is a negative one, there could be a huge psychological price for you to pay. Clients describe the feeling as letting their dog down at the time he needed them the most (and guilt is usually a really tough thing to work through).

Third, the veterinarian is often asked to be the decision-maker. We commonly hear, "What would you do if he was your dog?" Some vets don't hesitate to answer the question directly, welcoming the opportunity to tell their client exactly what they believe. (After such a response, it can in fact be difficult for some clients to have the gumption to choose something different!) When I'm asked the question, I am honest with my answer, but I issue it with the disclaimer that what feels right for me and my dog may not be the best choice for my client and patient. Truth be told, when it comes to truly difficult decisions, I'm often just as uncertain as anybody else. Think about whether you are asking your vet to make the decision for you, or simply trying to gain perspective.

The fourth option, and my obvious favorite, is what I refer to as (drum roll here) the "responsible advocate" approach. No, it's not nearly as easy as the other three choices because some serious research and soul-searching are required. But the payoff is huge. You will end up making the decision most likely to serve your dog's best interest, and no matter the outcome, you are assured of having the greatest peace of mind. Just keep on reading—the remainder of this chapter is loaded with advice to help you when faced with tough choices.

SECRET FOR SUCCESS

Anthropomorphism Not Allowed!

To anthropomorphize literally means to attribute human form or personality to things not human, which includes assuming animals have the same feelings or behaviors we do. Anthropomorphizing is a fact of life for most of us animal lovers—it's so darned enjoyable. (I don't dare give you details of what goes on in my household!) So, go ahead and humanize your dog to your heart's content. Let your dog join you at the dinner table, dress him up for Halloween, ask him to be the ring bearer at your wedding. But please, do not let your anthropomorphic tendencies get in the way of what's best for his health.

How is this relevant? Consider the following example. I perform oodles of ultrasound evaluations in my day-to-day practice. It is a fabulous diagnostic tool, because it's a comfortable, noninvasive, and risk-free procedure for the patient, and provides a wealth of useful information to the veterinarian. In order to get good ultrasound images, the body part must be shaved—for example, to view the abdomen, it's necessary to shave the underside of the dog's belly. Every so often, I encounter a client who declines the ultrasound because of her certainty that a bald belly would result in a horrific case of canine embarrassment, resentment, or depression (yes, I've heard all of these reasons). This is anthropomorphism at its best, or perhaps I should say at its worst.

Don't assume that your dog will be miserable and his life won't be worth living if he loses his eyesight. The majority of dogs adapt to blindness remarkably quickly and completely. In fact, sometimes I am the first one to recognize that my patient is blind. He has been moving about so well in his home environment that his human family didn't even notice. And, the same can be said of limb amputation. Most dogs, typically, are back to their usual tricks within a few weeks. They don't worry about what their buddies at the dog park will think when they see three legs rather than four.

The bottom line is that anthropomorphism may cause you to dismiss legitimate options in your dog's medical care. I encourage you to set aside your preconceived notions based on human responses and experiences, so you can really hear what your vet has to say.

Your Dog's Best Interest

When I counsel people who are working through tough medical decisions for their dogs, I commonly hear, "I want to be sure that I'm not just doing this for me." I view this comment as a very good sign, because it means the person is trying to stay focused on their dog's best interest rather than their own needs and desires. But, how do you know what's truly right for your dog?

What Are the Facts?

I have nothing against good old-fashioned "gut" feelings, but when it comes to your dog's health, I recommend being well-informed over "going on instinct" any day. For every health issue, there will be plenty of information into which you can sink your teeth. I encourage you to take a really big bite: read all you can, and ask your vet to explain the pros and cons of every diagnostic and therapeutic option. Your goal is to gather and process as much knowledge as possible in order to make an informed choice (see earlier discussion on p. 10).

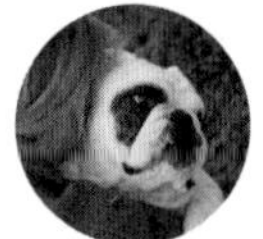

74% of dog owners consider their pup a "child" or family member.

Does the Medical Plan Make Sense?

Should you give your dog antibiotics when there's no evidence of a bacterial infection? Perform a diagnostic test when it's not clear how the results might change what happens next? Vaccinate for "the works" without knowing which diseases he might be exposed to?

Asked this way, these questions almost seem silly, but I must tell you that, as an internist who receives referrals and provides second opinions, I encounter situations where testing was performed or treatment administered that truly didn't make sense in the case at hand. If your veterinarian is prescribing care that, in your mind, doesn't "jive" with your dog's problem, I encourage you to question, investigate, poke, and prod, until things "click."

I know how difficult this can be—heck, I know how hard I have to work to avoid glazing over when listening to my insurance broker explain unfamiliar things such as copays, benefits, indemnities, and liabilities. However, it is in your dog's best interest to ask to see his lab test results, ultrasound images, and X-rays rather than just hear about them. It is with his well-being in mind that you should ask for clarification when your vet summarizes a complicated procedure. You are not being a nuisance, and a good veterinarian will not be offended by your questions (none of us are perfect, as I explain in chapter 4). Think of it as a healthy system of checks and balances.

In fact, most vets derive satisfaction from knowing that their patient's course of treatment makes sense to their client. They know that the more their client understands, the better their patient will be served.

Most vets derive satisfaction from knowing that their patient's course of treatment makes sense to their client.

Does Your Dog's Personality Play a Role?

Some dogs are fighters, and others are wimps; some are shy or fearful with people they don't know well, and others love every opportunity to lick a new face; some have high pain thresholds, while others "scream" when their temperature is taken. Some love being at the vet's because it's an opportunity to get cookies and attention. Others rebuff all versions of vet-clinic bribery. Your dog's personality is an exceptionally important factor to consider when being his advocate.

Consider the following example. In my practice, I work with many dogs with cancer and chemotherapy is a common topic of discussion.

When I counsel people on its pros and cons, I encourage them to think about their dog's emotional response to being taken back and forth to the veterinary clinic for frequent visits.

Honest to goodness, some dogs absolutely love coming in for their chemo: they get lots of hands-on attention from people who adore them, and in my clinic, there is a 10-cookie minimum per visit. On the other hand, there are the dogs that learn to hate getting in the car because they think it means another trip to the vet. These dogs shake, quiver, and worry from the moment they leave home until the moment they get out of the vet's office. And, they can be starving, but still refuse to eat a cookie in our presence.

Your dog's personality is an exceptionally important factor when being his advocate.

So, it's pretty easy to figure out which personality type is better suited to the rigors of chemotherapy. If a client is uncertain, I generally recommend proceeding, and reevaluating that decision should the dog protest.

I must admonish you here; please don't assume your dog's emotional response is the same as yours. Based on personal experience, you may be aghast at the thought of surgery. Projecting this onto your dog, though, might steer you toward a decision that is not truly in his best interest. Review the earlier sidebar called "Anthropomorphism Not Allowed!" to help you stay focused on your dog's inherent personality traits and emotional responses rather than your own.

What Is Your Dog's Job Description?

When I ask about your dog's job, I really want to know what your dog loves to do more than anything else in life. Perhaps he is a Border Collie whose passion is running all day to herd and keep track of everything that moves on his property. Or, maybe he's a lapdog and his most favorite thing in life is snuggling with you on the sofa. You want to be sure the choice you are considering doesn't interfere with his ability to continue doing his favorite things. For example, successfully treating a dog with heart failure requires significant lifelong exercise restriction. This might be no problem for the couch potato but perhaps very hard to endure for the Border Collie.

You want to be sure the choice you are considering for your dog does not interfere with his ability to do his favorite things in life.

How Does Your Dog's Age and Overall Health Play a Role?

I often talk with clients about whether or not to proceed with significant testing, perform a complicated procedure, or start an involved treatment plan for their older dog. They commonly voice concerns based on his age. I might hear, "I'd say, 'Yes,' if he wasn't already 12 years old." Sounds logical, doesn't it—at least initially? However, when you stop to think about it (and you can bet that I encourage my clients to think about it), you realize such a knee-jerk response doesn't necessarily add up. It might not make sense to surgically repair the torn knee ligament in a sedentary eight-year-old dog that is already debilitated by arthritis, but it may very well be a good idea to perform the same surgery on a 12-year-old that cavorts daily at the dog park.

You see, when it comes to making decisions for your older dog, consider his functional, not chronological age. (This has nothing to do with multiplying or dividing by seven!) So, rather than nixing the notion of a test or treatment based on a number, make your decision based on the quality of your dog's life before he became sick or injured.

57% of people would prefer their dog as their only companion if they were stranded on a desert island.

Can You Change Your Decision if You Don't Like What You See?

By the time one of our own dogs, Lexie (a Border-Collie-mix, we think) was 17 years of age, she had arthritis in just about every joint of her dear little body. My husband and I decided to have a rehabilitation specialist work with her on an underwater treadmill. I've seen this produce utterly amazing results for arthritic dogs. Both times Lexie tried the treadmill, however, she told us in no uncertain terms that she disliked the whole experience immensely. No problem. We simply discontinued this therapy. Sometimes the knowledge that a change of heart will be perfectly okay turns a monumental decision into a much easier one.

Sometimes the knowledge that a change of heart will be perfectly okay turns a monumental decision into a much easier one.

What Does Your Dog Think?

I encourage you to spend a little nose-to-nose and eyeball-to-eyeball time with your dog. Despite his not feeling "up to speed," does he still have that recognizable spark, or maybe a twinkle in his eye to indicate he wants to keep on fighting? Do his eyes look tired in a different way? I am firm in my belief that such one-on-one time can provide invaluable feedback about what your dog would have you choose on his behalf. And, nothing is more important than that.

What Best Serves YOUR Peace of Mind?

By virtue of the fact that you are reading this book, I know you care deeply for your dog and want to act in his best interest. What's best for *you* is also important. I'm not necessarily talking about peace of mind in the near future. Rather, I'm referring to the kind you desire after having lived with your decision weeks, months, or even years down the

SECRET FOR SUCCESS

Get Comfortable with the Concept of Benign Neglect

It's known that approximately 80 percent of maladies seen by the average human physician fully resolve without any medical intervention. No such estimate exists in veterinary medicine, but I suspect that the findings would be similar. Many people feel that a trip to the veterinarian with a sick dog is unrewarded if they don't return home with medication in hand. In fact, vets often feel pressured to prescribe something, even when "nothing" would be best.

When "benign neglect" is recommended, consider yourself lucky! Your dog's illness is not serious and you don't have to hassle with medication. As the saying goes, "Don't just do something, stand there."

road. Canceling a long-awaited holiday in Hawaii because your dog needs surgery may be very disappointing at the time, but in the long run, it may be the decision you can best live with.

Finding peace of mind is the natural consequence of stepping up to the plate as your dog's responsible medical advocate. This doesn't mean that you will invariably choose the most aggressive option—rather, your choices will be guided by what you ultimately believe are the right ones for your dog. Even if the end result isn't what you'd hoped for, you will be at ease knowing that your actions were always guided by your good intentions.

Finding peace of mind is the natural consequence of stepping up to the plate as your dog's responsible medical advocate.

No matter how refined your medical advocacy skills are, some circumstances render it impossible to make decisions based solely on your dog's best interest. This can happen when financial issues come into play, or when it's simply impossible for anyone to know which choice is truly best. On the following pages I provide some exercises to help you deal constructively with these challenges.

When Cost Gets in the Way

I suggest the following exercise when clients are in the throes of decision-making and I see that concerns about money are taking center stage. I ask them to imagine that they are wealthy beyond compare. Once they've envisioned themselves as Bill Gates or Oprah Winfrey, I ask them to have another look at the choices on the table. Once cost is removed from the equation, it's often much easier to hone in on the decision that truly serves the dog's best interest.

Better yet, sometimes there's the added bonus of realizing the best option is the less expensive one. But, what happens when the obvious best choice is not affordable? Rarely is there only one right way to do things, so the key to making things work in this situation is examination of medical alternatives and different ways to pay the bill. Refer to chapter 10 for more coaching about money matters.

When No One Knows What's Truly Best for Your Dog

In most situations, the client can be assured peace of mind based on the knowledge that she acted in her dog's best interest. However, sometimes medical circumstances are such that it is simply impossible to know which choice serves this purpose. Consider the example of my dog, Lily.

Approximately 20 years ago, my husband (also a veterinarian), young children, and I shared our lives with a gentle, loving Golden Retriever named Lily. She was an incredibly hale and hearty dog, even as she entered her "golden oldie" years.

Shortly after her thirteenth birthday, we noticed our Lily girl slowing down in the food department. Rather than wolfing down her meal within seconds, she dawdled a bit at the bowl, sometimes even leaving a small amount uneaten. If you are familiar with this breed, you know this is a serious symptom—in fact, most Goldens would quit breathing before they'd stop eating!

An abdominal ultrasound showed that Lily had a large mass within her liver. We obtained a biopsy using a nonsurgical ultrasound-guided technique, but the pathologist couldn't be certain whether the mass was benign or malignant. My family and I were faced with a monumentally difficult decision. Dare we ask our beloved 13-year-old to endure major surgery in the hopes that the mass was benign and could be successfully removed? Or, should we simply provide her with supportive care, keeping her as comfortable as possible for as long as possible?

Surgical removal of a liver lobe is no walk in the park for any dog, even one much younger than Lily. What if her life ended as a result of surgical complications? How would we feel if the mass turned out to be cancerous and couldn't be removed? How would we live with ourselves knowing that we'd subjected our dear dog to such an invasive procedure during the last few weeks or months of her life?

We thrashed around with our decision for days, looking at it from many different angles, and discussing all the pros and cons. We spent time, lots and lots of time, observing and "talking with" Lily. We studied her demeanor and expression—especially what her eyes were telling us. Did we sense that she was ready to "throw in the towel," or was she game to take that giant step into the surgery suite?

No one could predict whether surgery would do her more harm than good. When such uncertainty exists, I encourage you to work through what I refer to as a "peace of mind exercise." My husband and I used this exercise to make our decision for Lily.

Here's how it worked in our case: we were considering two options. The first was surgery in the hopes of successfully removing her liver tumor. The second was to forego surgery, and use supportive measures (special diet, antinausea and pain medication) to keep Lily as comfortable as possible, for as long as possible. Although we felt, based on her overall health, Lily was a reasonable surgical candidate, we had no way of knowing which of these two options was truly in her best interest.

The first step in the "peace of mind exercise" is to play every option out to both its negative and positive conclusion. In other words, decide what the best possible outcome and the worst possible outcome are for each. When my husband and I did this, we came up with the following:

1. If we opted for surgery, the best case would be that her disease would be cured and her normal good quality of life restored. In the worst case, the surgeon would be unable to remove the tumor, and additionally, there would be significant surgical complications, perhaps resulting in death.

2. If we went with the supportive-care option, the best scenario—even if the mass was benign—would be that Lily's symptoms would progress slowly, and she would have another few months of reasonably good quality time. The worst imagined outcome would be rapid progression of her symptoms resulting in the need to consider euthanasia within a few weeks.

Step two of this exercise is to determine which set of outcomes would best serve your peace of mind. With Lily, our goal wasn't to find an option we "liked" (impossible under the circumstances), but rather to determine which we could most readily live with. This was the key to making our choice about how to proceed.

My husband and I were fortunate in that we found ourselves on the same page—not always the case when more than one decision-maker is involved. The process certainly wasn't an easy one. It took a great deal of thought, investigation, introspection, and yes, it required "discussions" with our dear old dog. When we made our choice, did we know with certainty that it was correct? Absolutely not! We did know it was well-informed with nothing but the best of intentions for our sweet girl. No matter how things turned out, we doubted we would have regret; sadness and disappointment perhaps, but no regret.

So, what ever happened to Lily? We decided to take the more aggressive approach. We asked a board certified surgeon to attempt to remove her liver mass. The surgery lasted almost four hours, and thank goodness, the mass was removed in its entirety. And, it turned out to be benign rather than cancerous—the icing on the cake! She took a considerable amount of time to completely recover, but within a few weeks we had our Lily girl back. Her next three years were spent in Golden Retriever bliss (good food, good company, and the opportunity to swim on a regular basis).

Lily lived to the ripe old age of 16! We felt extremely fortunate with the outcome of our decision, but knew in our heart of hearts, that had surgery not turned out well, we would have had peace of mind knowing that we'd done our very best, and stayed true to our good intentions throughout the decision-making process.

QUICK REFERENCE

Questions to Ask When Determining What's Best for Your Dog

- What are the facts?
- Does everything make sense?
- How does your dog's personality play a role?
- What is your dog's job description?
- How do age and overall health play a role?
- Can you change your mind if you don't like what you see?
- What do you think your dog is telling you?

“Never go to a doctor whose office plants have died.”

—ERMA BOMBECK

3

Finding Dr. Wonderful and Your Mutt's Mayo Clinic

When it comes to choosing a veterinarian and clinic for your dog, your choices will be somewhat dependent on where you live. In some rural locales, there may be only one vet who practices on all creatures, large and small, within 100 miles. In the midst of a large city, more than a dozen veterinarians might be available within a 5-mile radius. There may be vets who practice on their own, or those who run a clinic as part of a group. There are general practice veterinarians (what we would call "family practitioners" in human medicine) who are responsible for routine care, and specialists, such as cardiologists, neurologists, and dermatologists, to whom you can be referred (see more about this on p. 79).

Some vet clinics are devoted to small animals only, while others extend care to larger species (mixed animal practice). In places farther flung, the "local" veterinarian's office might do everything, from "in-and-out" checkups to major surgery requiring an overnight stay; in urban areas, it is more common to find separate hospital facilities devoted to 24-hour care, specialist referrals, and emergencies. Veterinary schools often have "teaching hospitals," which have ready access to the most current diagnostic and therapeutic technology and

QUICK REFERENCE

Veterinary Medical Teaching Hospitals

A veterinary medical teaching hospital houses board certified specialists who are training students, interns, and residents. Vet schools are the classic example, but some private specialty practices also have active teaching programs.

The upside to seeing a vet at a teaching hospital is that the instructors are typically clinicians with a good deal of smarts who are involved in research pertaining to various specialties and "in the know" about all that is current in their area. These hospitals often have ready access to the most current diagnostic and therapeutic technology.

The downsides are that your dog's evaluation will be a more labor-intensive process: a vet student may be the first person to evaluate your dog, followed by an intern or resident, and finally a clinician (instructor). This is all fine if your dog loves people and you have a good deal of time on your hands!

a reputation for charging less money for their services (see sidebar).

It should now be apparent you face an array of options when choosing a veterinarian and clinic for your dog. And, regardless of where you live, you have the right to be fussy. If you are not satisfied with the "only show in town" it simply means you'll need to do some extra driving to take care of your dog's health care needs.

So, which is more important, choosing the right veterinarian or the right veterinary clinic? Ideally, you want the whole package and, the good news is, they often come hand in hand. Veterinarians who want to do a great job don't last long in crummy offices. Likewise, sub-par vets don't "cut the mustard" in facilities that provide first-rate service. Might a little bit of compromise be reasonable when making your choice? Absolutely—as long as you feel confident that your dog's health will not suffer for it.

Dog people tend to believe—at least initially—that all veterinarians are *wonderful*! After all, vets must love animals, and we have to be very, very smart (everyone knows how hard it is to get into veterinary school). Interestingly, people seem to be far less skeptical of their veterinarian's capabilities and intentions than they are of their own physician's. Speaking as a vet, I wish we all deserved such extraordinary benefit of the doubt, but that just isn't the case. Official veterinary disciplinary organizations exist for a reason. I certainly had a few classmates I wouldn't let near my own sick dog with a 10-foot syringe (then *or* now).

Veterinary clinics are not all created equal. A front office may appear clean and inviting, but enter its "bowels"—the place your dog gets to see when they take him to the "back"—and you might find an environment that is far less appealing. I know it's most convenient to go to the clinic right around the corner, and perhaps it *is* just the right fit. However, some research is necessary before coming to that conclusion.

Getting Started

There are more than 74 million pet dogs in the US.

How do you go about finding the ideal family veterinarian who works in a great clinic? Begin your search by asking friends and neighbors. If they are passionate in a particularly negative or positive way, learn why. Their reasoning may not be relevant in your case—your neighbor may be excited about the "free vaccine with every office visit" policy, whereas such an enticement might not appeal to you. It may even be a turnoff.

Next, "shmooze" with others at your community dog park, and see if there's a negative or positive consensus regarding the vets in your area. Talk to dog trainers, and the staff at pet stores, feed stores, grooming parlors, boarding facilities, and your local humane society. Ask them who they've chosen for their own dog's care. If the same names keep popping up—whether in association with complaints or compliments—take note. Such information is likely quite reliable.

Although not necessarily easy to obtain, far and away the most trustworthy and accurate opinions come from staff at your local emergency hospital. Many family vets refer their after-hour emergencies and patients requiring overnight care after a procedure to such a facility, and people working there readily acquire information (and form opinions) about the referring veterinarians—how they think and their ability to make important medical decisions. Hospital staff interact directly with these vets, read their medical records, and observe the quality of their X-rays, surgical techniques, bandages, splints, and anesthetic recoveries. They also get a clear sense of clients' level of satisfaction with the care their dog has received. Trust me—if a client is unhappy, the emergency hospital staff members are going to hear all about it!

The most trustworthy and accurate opinions come from staff at your local emergency hospital.

So how do *you* get the "lowdown"? This can be a little bit tricky as staff may be reluctant to share their recommendations. After all, their business depends on referrals from general practitioners within the community. If favoritism is perceived, the hospital's business will suffer. Here I've offered some advice (I hope you don't mind the little bit of bribery involved!)

Pay your local emergency facility a visit, ideally during a quiet time—call ahead to see if the waiting room is somewhat empty. Arrive with your arms filled with home-baked goodies. Give the receptionist a sense of what you are after, and ask to speak with the emergency vet, promising that you only need two minutes of her time.

If allowed access, try to have a private conversation and get straight to the point. Bend over backward to reassure the vet that you will be discreet. Say that you are looking for exceptional medical care for your dog, and ask for a few (not just one) recommendations. This way, she can feel she is not showing favoritism. You can also provide her with a list of names already suggested to you. Listen carefully and watch the doctor's response—she might be reluctant to give you an out-and-out negative opinion, but her body language may tell you a great deal.

Don't forget to also tactfully solicit opinions from other hospital staff. Although technicians and receptionists are not always forthcoming with recommendations, when they are, count your blessings. You've received some exceptionally valuable information.

In Search of the Ideal Veterinarian

The fact of the matter is veterinarians have different educational backgrounds, and diverse personalities, egos, and levels of expertise. One may be brilliant in the operating room but insecure when speaking with clients in the exam room. Another may have an extraordinary "bedside manner" but be deficient in medical reasoning skills. Some vets are humble, and others have an ego the size of a colossal manatee.

And then, there are those who are the "whole package." No, they are not perfect, but they are compassionate people who love working with dogs and their humans. They are practical-minded, have proficient medical and surgical skills, and are willing and able to recognize when

SECRET FOR SUCCESS

Check with the Authorities

During your search for the ideal family veterinarian, it is reasonable to contact your state veterinary licensing board to help weed out any "bad apples." Know, however, that it takes some seriously bad behavior to appear on a board's "Most Wanted" list! Don't assume that someone with a clean state record is an exemplary veterinarian.

something is beyond their capabilities. It will take some work on your part, but it is perfectly reasonable to expect to find this "ideal."

Figure Out What's Important to You

As you embark upon your search, create your list of important qualities:

- **Do you want a young vet with enthusiasm who is well versed in cutting-edge technology? Or, a seasoned veteran who has amassed a great deal of experience and intuition?**

- **Do you want to work with a "mixed" (large and small) animal practitioner, or someone who specializes in dogs and cats?**

- **Does your pup prefer women to men? (Truth be told, when dog's do have a gender preference, it's invariably for women vets—sorry guys!)**

- **Are you a better fit with a doctor who offers forth every available option, or one who makes a strong recommendation?**

- **Do you prefer a more aggressive or more conservative diagnostic and treatment approach?**

- **Are you interested in holistic options rather than an exclusively Western or Eastern medical approach?**

- **Does your dog have special needs? For example, are you planning to breed your dog and therefore want a veterinarian with expertise in reproductive medicine?**

Of course, it's possible that you may have two dogs with distinctly different medical needs, in which case you should make a list of important qualities for each one's ideal vet. And believe it or not, you very well may need more than one doc on your health-care team!

The Importance of Communication

It's a safe bet you want your dog's doctor to be technically skilled and an astute diagnostician. But have you considered the possibility that her communication style may have a significant affect on your dog's health? Research in human medicine has clearly demonstrated that the manner in which physicians talk and listen has a direct impact upon health outcomes. The more the patient understands and is involved in decisions, the greater the likelihood of a positive outcome. Although such a study has yet to be performed in the veterinary profession, after almost three decades of living and breathing veterinary medicine, I cannot help but believe the findings would be similar.

Paternalistic and Relationship-Centered Care

In the world of human medicine, two major styles of communication exist between physicians and their patients. One of these models is referred to as "paternalistic care"—when the doctor does the majority of the talking. The importance of the patient's opinion and feelings is minimized. In fact, they may not even be solicited. It's all rather business-like with little room for small talk: the doctor sets the agenda, and in lieu of explanation and discussion of options, she counsels the patient on what she deems to be the best course of action.

The more the client understands and is involved in decisions, the greater the likelihood of a positive outcome for her dog.

The second style is referred to as "relationship-centered care." In this model, explanations and options are provided. The doctor uses an individualized approach and takes into account the patient's feelings and lifestyle, so a sense of collaboration exists in determining what is in the patient's best interest.

Veterinary Communication Styles

At the time of publication, only a tiny amount of research had been done evaluating communication between veterinarians and their clients. What was learned is, just as in human medicine, there are paternalistic and relationship-centered styles. Some vets clearly incorporate one model only, whereas many utilize a combination of the two.

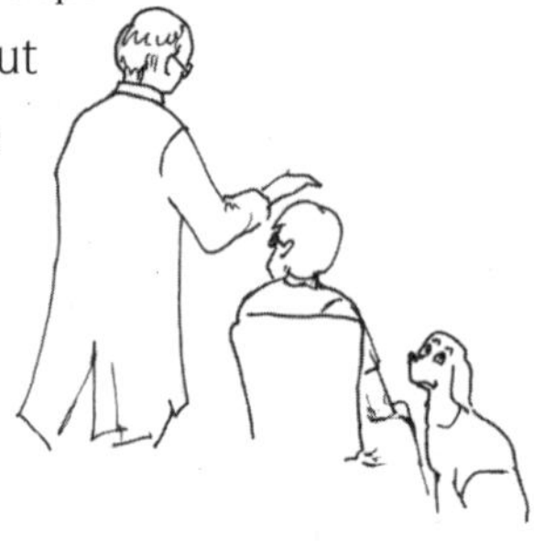

In the paternalistic approach, the veterinarian keeps an emotional distance with no discussion about personal matters, and little is learned about the role the dog plays in the family. Such a vet voices what she perceives is best for the patient before her, without considering the patient and client's unique situation. For example, she might say, "Take him to surgery. That's what I would do if he were my dog," without first gaining a sense of the client's feelings toward surgical procedures for the medical issue at hand. Or simply, "Give him these pills three times a day, and call me next week if he isn't any better," without first determining if three-times-a-day administration is possible.

More than 13 million Americans celebrate their dog's birthday.

In the relationship-centered care model, the client's opinion and feelings are held in high regard and enough time is allowed during the office visit to hear them. The vet takes the time and effort to recognize the special or unique role the dog plays in his human's life. She has a strong sense of the level of emotional attachment and recognizes that her responsibilities expand beyond the patient to include the well-being of the client. Empathy and support are provided. Instead of *telling* the client the dog needs surgery, she may *suggest* it and ask if the client has concerns or questions; and instead of simply prescribing an antibiotic that needs to be given three times a day, the veterinarian might first ask the client if she can manage such frequent administration.

You Have a Choice

It was no great surprise (at least to me) that when the two styles of human physician/patient interactions are compared, the relationship-centered model clearly results in more favorable health outcomes: for example, better diabetic regulation, improved pain management, and lower blood pressure measurements. As I mentioned earlier, little research has been done on the veterinarian/client/patient scenario, but as a vet who leans toward relationship-centered care in my own practice, it is my belief that the findings would likely be similar.

SECRET FOR SUCCESS

Make the Veterinary Staff Part of Your Extended Family

People do not work at veterinary clinics to become rich or famous, but because they care deeply for their patients. That feeling often spills over to their clients—the advocates. I encourage you to befriend the veterinary staff, and bend over backward to show your appreciation. What are the chances that your sick dog can be squeezed into an already busy schedule on a moment's notice? Without doubt, they'll be greater if the person answering the phone knows and likes you.

Despite this, there are certainly people who prefer a vet who favors the more paternalistic style. So, I encourage you to settle for nothing other than your first choice, whichever it may be.

Time to Pay a Visit

A veterinarian expects to be interviewed by a prospective client, and I encourage you to do exactly that. This is an important step in the process of determining who gets the "honor" of becoming your dog's doctor. In addition to speaking directly with the veterinarian, plan time to interact with the veterinary staff and take a tour of the clinic (see more about this on p. 53). By all means, bring your pup with you. After all, his opinion is important, and he may just get a treat or two for his time and trouble.

Try to get a sense of whether or not the prospective vet enjoys her work. Had I to choose a profession all over again, I wouldn't change a thing. I get to interact with fabulous dogs and their devoted humans

SECRET FOR SUCCESS

Spend Some Time "Schmoozing" with the Staff

Spend some time getting to know the staff. Learn their names, and ask them about their own pets and how they came to work in a veterinary office. Acknowledge those who do their job well, and be sure that their supervisors know of your positive impressions. Demonstrate your appreciation verbally, and with thank you cards or letters—and don't forget to include a photo of your pup! Bake cookies, bring in flowers, or have your child draw a picture for the vet and her staff. This won't be considered bribery. Rather you will be seen as a client who truly values the work they do.

all day long. The medicine is stimulating, the people are fascinating, and the work is never dull. However, some vets don't necessarily feel the same—after all, there's also a business to run, employees to manage, a physical plant to maintain, long hours, difficult clients, and too many euthanasia procedures. Some suffer from job fatigue or emotional burnout, and become disgruntled, unhappy people. If you sense this "vibe," best to continue your search for a veterinarian with more positive energy.

Consider the list of suggested questions in the sidebar on p. 50 before you pay your visit, or even take them with you. The correct answer to most of your questions is the one that "feels right" to you. In addition, consult the list of "deal breakers" on p. 57. (Note: this list applies to both your search for the right vet *and* the search for the right clinic, which I'll discuss next.) These are the issues with no room for compromise. I encourage you to add questions and "deal breakers" of your own based on what will be best for you and that wonderful dog of yours.

QUICK REFERENCE

Questions to Ask when Interviewing a Prospective Vet

❶ **Does the veterinarian practice Western medicine, Eastern medicine, or holistic medicine (a combination of both)?** You probably know what type you prefer, so find someone whose expertise matches your needs.

❷ **Is her bedside manner well suited to you AND your dog?** Does she greet both of you when she comes into the room? Is she outgoing and personable, or businesslike? Is her demeanor overwhelming or too reserved for your dog? Does she make eye contact with you? Does she seem rushed or relaxed? When you ask a question, does she give you too much, too little, or just the right amount of information?

❸ **From which school did she graduate and when?** This information may be important in a couple of ways. First, graduation from a highly regarded veterinary school gives you more certainty about the quality of her education (whether or not the student availed themselves of such an educational opportunity is anyone's guess). Second, knowing "when" gives you the opportunity to decide if you want to work with a younger vet who has been trained in the latest procedures and gadgetry, or a seasoned vet with loads of experiential wisdom.

❹ **Has she received advanced training?** After graduation from vet school, some opt for an internship with or without residency training rather than going straight into private practice. This provides an additional one to three years of hands-on experience in a teaching hospital under the guidance of specialists in medicine and surgery (see p. 79 to learn more about veterinary specialists).

❺ **Is she a solo practitioner or part of a group?** Working in a group setting means that a vet can readily share ideas and review tough cases with her colleagues—a huge advantage. The drawback of a group practice is that you might encounter the "multiple doctor dilemma" that interrupts continuity of care. It is worth asking if it will be possible to regularly see the vet you are interviewing (see more about group practices on p. 39).

❻ **How does she handle after-hours emergencies? If referred, where and to whom?** It may feel reassuring to hear that the veterinarian takes all of her own emergency calls, but this can be a mixed blessing. It's natural to always want access to the doc who knows your pup so well, but any vet who handles a fulltime caseload plus emergencies is (or is about to become) burned out.

❼ **How does she handle sick patients requiring 24-hour care?** Does the vet refer to an emergency hospital (see p. 39) that is open evenings and weekends? If patients remain in her clinic, who cares for them throughout the night? I wish it weren't so, but leaving patients alone

Questions to Ask when Interviewing a Prospective Vet cont.

from six o'clock at night until six o'clock the following morning is the norm in some clinics.

❽ **Whom does she ask for advice and second opinions?** Does the prospective vet have a working relationship with veterinary college faculty members or veterinary specialists in private practice? Does she have access to a veterinary radiologist to review X-rays? Does she solicit opinions from veterinary specialists online? If none of these, consider this a red flag.

❾ **Does she have a special interest in any particular area of medicine? What types of medical or surgical cases does she feel uncomfortable with and would refer elsewhere?** If the vet is an expert in the field of hormonal imbalances and your dog has diabetes, what a fabulous fit! By the same token, if a vet tells you she's not crazy about reproductive medicine, and you're having trouble getting your Bassett Hound pregnant, perhaps you should keep looking. Note: if your prospective doc tells you she's comfortable with absolutely everything, buyer beware! I've never met a veterinarian who has expertise in every aspect of canine medicine and surgery. A 100-percent-do-it-your-selfer may be hazardous to your dog's health.

❿ **Is she a member of any professional organizations? Which ones?** Most veterinarians are dues-paying members of at least a couple of veterinary associations, whether national, state, or regional. The absence of professional affiliations suggests that the individual standing before you may not be keeping pace with her profession.

⓫ **What type of continuing education does she receive? How often?** Given how rapidly veterinary science changes, it is impossible to stay current without continuing education on a regular basis. Some states mandate a minimum number of credits per year.

⓬ **What is her philosophy about routine vaccinations? Does she follow American Animal Hospital Association (AAHA) guidelines?** Please read through chapter 6 before considering the response to this question. We know a great deal more about vaccines than we did five years ago. Routine yearly immunizations are no longer considered to be in a dog's best interest. Any veterinarian still recommending these or the same "vaccination package" for every dog is likely unaware of or ignoring current research.

⓭ **Does she make house calls?** For some dogs, riding in a car is a terrifying ordeal. Or, perhaps you rely on public transportation and travel is complicated. And there are those pups that are traumatized by the vet clinic environment. Some vets see patients as house calls, as well as at their clinic, which might be just the right choice in these cases. Note: If you choose a vet who only performs house calls, find out where procedures such as X-rays or surgery are performed.

Questions to Ask when Interviewing a Prospective Vet cont.

⓮ **Does she use pain medication when spaying and neutering?** I'm embarrassed to say that when I graduated from veterinary school, these so-called "routine" operations were performed with no postoperative analgesia. Thankfully, the standard today is to provide pain meds with surgical procedures. How this question is answered will give you some insight about whether or not she is practicing state-of-the-art medicine.

⓯ **Does she perform tail-docking and ear-cropping?** These are surgical cosmetic procedures performed on puppies so they will conform to show-ring breed standards. For example, an Airedale Terrier's tail is cut short and a Doberman Pinscher's naturally floppy ears are cropped to achieve a pointed upright look. The veterinarian's philosophy about this controversial topic may be important to you.

⓰ **How does she handle euthanasia?** No one wants to think about this question, but the unfortunate likelihood is that, at some point, you will be forced to face this option for a beloved dog. Will the vet make a house call for this procedure? Does she allow or encourage family members to be present? Does she schedule the procedure for the end of her workday, so that she won't be pushed for time? Can she accommodate it at any time when an urgent situation develops? For help in interpreting the answers to this question, please refer to chapter 11.

⓱ **Is she planning to leave the community or retire in the near future?** It is frustrating to put energy into your selection only to find out that your number-one choice is moving to New Zealand in three months time. So ask! You want Dr. Wonderful to be "tied" to the area, in love with her job, and intending to continue working even when she wins the lottery. (Note: if she has a few kids who are college-bound, she likely won't be retiring anytime soon.)

⓲ **Does she have a dog of her own?** Although the answer to this question may be completely unimportant, go ahead and ask it! You're likely to receive a big smile and hear a good story or two.

Finding Your Mutt's Mayo Clinic

Finding Dr. Wonderful is only part of the equation. Many people neglect to carefully examine the veterinary clinic where she works—and it deserves as much attention as her credentials. Keep in mind that a veterinarian cannot do her job well without a first-rate physical plant and exceptional support from her receptionists, technicians, assistants, and housekeeping staff.

A veterinarian cannot do her job well without a first-rate physical plant and exceptional support from her receptionists, technicians, assistants, and housekeeping staff.

Consider the clinic itself. It's great if you find the waiting room and exam room to be clean, comfortable, and appealing, but they likely occupy only a small part of the overall floor plan. Ask to view all the rooms your dog gets to see when he is taken "to the back."

- The cages or runs should be clean, well ventilated, and roomy. Your 140-pound Great Dane can't possibly be expected to recuperate from an illness in a space that is only big enough for a Labrador Retriever.

- The surgery suite should be devoid of almost everything except a surgery table, instrument stand, and light fixtures. A sterile surgical procedure certainly can't be performed in a room that doubles as a storage closet.

- The treatment room is the busy hub of the hospital. This is where intravenous catheters are placed, dental procedures are performed, and critical patients are hospitalized. It is also where you want your dog to recover from anesthesia, not back in the kennel area where a problem might go unnoticed.

NO NO BAD DOG Sometimes unusual factors—things we'd never predict—come into play when choosing a veterinary clinic. I'll never forget No No Bad Dog, an adorable mutt who appeared to be one-part poodle and who knows what else. No No's human counterpart was enthusiastic about bringing him to see us because of the glowing recommendations she'd received about our clinic and staff. But, the very first time No No appeared in our waiting room, he surprised us all by having a full-blown seizure. He was fine after a minute or two, and his rather startled companion told us that she'd never witnessed anything like it.

Well, wouldn't you know it, the next time he entered our clinic for a routine appointment, seizure number two occurred. Clearly, something—perhaps a particular smell or site—was causing the seizure. This was new to us, and I have never seen anything quite like it since. I had to suggest they try another clinic, and last I heard, the pup never experienced another seizure. Our place and No No Bad Dog simply weren't a good fit.

- The clinic should maintain an isolation ward for those dogs with potentially contagious diseases. It would be a shame for your pup to go to the vet's with one ailment only to leave with another.

The veterinary staff play a major role in the medical care process. These are the people sterilizing the surgical instruments, drawing up vaccinations, monitoring anesthesia during surgery, filling prescriptions, preventing your dog from leaping off the exam room table, disinfecting the rooms, and determining whether or not they can squeeze your sick pup into an already busy appointment schedule. No matter how good the veterinarian may be, incompetent staff or those with negative attitudes can have a devastating impact on the quality of care you and your dog receive.

As when searching for the ideal veterinarian, I encourage you to use the questions I've provided in the sidebar on p. 55 to help you decide whether or not Dr. Wonderful's clinic rates as high in your esteem as she does.

QUICK REFERENCE

Questions to Ask when Considering a Vet's Clinic and Staff

❶ **Is the staff pleasant?** Do you and your pooch receive a warm and genuine greeting? Do various staff members initiate conversation as you tour the facility, or are you ignored? Do you observe smiles and lively conversation, or do you sense general unhappiness and gloominess? Keep in mind the receptionists, technicians, and custodial employees are all part of your dog's "health-care team." Even a top-notch vet cannot do a good job without her support staff.

❷ **Does the staff seem knowledgeable?** Ask a technician about heartworm preventative medication. Ask a receptionist about pet insurance. Are their responses forthcoming and accurate? If they don't know the answers, do they offer to find them for you?

❸ **What's the facility like?** Take a tour, keeping your eyes peeled and your ears open. Is it cluttered and crowded? Is it dirty or smelly, or could it win the *Good Housekeeping* seal of approval? Do patients have comfy looking bedding, or are they on newspaper? Are the cages or runs sufficiently roomy, or one-size-fits-all? Are they immaculate, or containing feces and urine? Is there a lot of whining and crying going on, or do the animals appear reasonably content? Are you allowed to see the entire clinic, or only a small portion? Where are patients taken for walks? In a fenced grassy area, or alongside a busy street? Is the atmosphere uptight, too relaxed, or just right?

❹ **Does the clinic have AAHA accreditation?** The American Animal Hospital Association offers accreditation rather like the certification process given to human hospitals. Only those clinics that meet a stringent list of requirements pertaining to physical plant, quality of care, diagnostic services, medical record keeping, and management can become accredited. Note: there are many outstanding veterinary facilities that have not gone through accreditation as it is not a mandatory process. AAHA accreditation simply means that some of your homework has already been done, and by an extremely conscientious organization. (For more information about AAHA accreditation, visit www.healthypet.com.)

❺ **What are the hours of operation?** When you work regular hours, the availability of evening or weekend appointments is an important consideration.

❻ **What is the office policy concerning emergencies?** When your dog requires urgent attention, will he be fit into the schedule (perhaps a number of appointment times are left open each day for just this purpose), or will you be referred elsewhere?

❼ **Are clients allowed to accompany their dogs to the "back" of the facility?** Often, dogs are whisked away to the "mysterious" back rooms of the clinic for various procedures. Unless you are a "fainter," you may wish to go along. Find out whether or not this is allowed.

Questions to Ask when Considering a Vet's Clinic and Staff cont.

❽ **What are clients in the waiting room saying to the staff and each other?** People are often quite vocal about expressing strong feelings of satisfaction or dissatisfaction. Spend some extra time observing various comings and goings. This is an opportune and perfectly appropriate time to be an eavesdropper!

❾ **How is the waiting room arranged?** Do dogs and cats wait for their appointments in the same area? If your dog goes berserk when a kitty is in sight, a common waiting room might be a huge negative.

❿ **How long are the office visits?** All other things being equal, if "Clinic Up the Street" has 20-minute office visits and "Clinic Down the Road" books 30-minute appointments, move on down the road. The longer the appointment, the more time for a thorough physical examination and discussion. Additionally, the overall pace of the office will feel more relaxed, which means the staff is likely to be less stressed.

⓫ **What is the visitation policy?** Are there specific visiting hours or time restrictions on visits for hospitalized patients? Can children (or even other dogs) visit? Is it allowed more than once a day? If visitation is not allowed, be wary. Such visits can be extremely important for the dog and the family's peace of mind.

⓬ **Does the clinic have an isolation ward?** As mentioned earlier, ideally, dogs with infectious diseases are housed well away from other patients, while remaining under the watchful eye of staff.

⓭ **What do you see in the operating room?** Remember, this should be as sterile an environment as possible, uncluttered and separated from the rest of the clinic and the outside world by closed doors and windows.

⓮ **Does the clinic employ registered veterinary technicians (RVTs)?** Veterinary technicians (nurses) are educated to perform their trade in a number of ways, from on-the-job training to college-level courses. In order to achieve RVT status, the individual must complete a number of rigorous courses, then pass a difficult examination. RVTs are licensed to perform higher level tasks than unregistered technicians. If a technician is wearing a nametag, look for the initials "RVT" after her name. Many very skilled techs are not RVTs, but having some RVTs on staff is generally a bonus in terms of the quality of care.

⓯ **What payment plans are offered?** Read chapter 10 to learn more about various ways to pay your bill.

⓰ **Is medical boarding allowed?** When your dog requires significant at-home medical care and you must travel, this is an important consideration.

Deal Breakers

- The vet vaccinates dogs for everything, every year.
- Adequate supervision and care is not provided overnight and on weekends.
- The vet is a "do-it-yourselfer." She doesn't solicit second opinions from specialists or refer difficult cases.
- Pain medication is not administered when spaying and neutering.
- The vet does not receive continuing education on a regular basis.
- The vet and her staff don't appear to enjoy their work.
- The staff isn't friendly or helpful.
- You are not offered a tour of the facility.
- The hospital is dirty or smells bad.
- Inpatients have dirty or inadequate bedding.
- Inpatients aren't allowed to have visitors.
- The vet and her staff use brute force (a.k.a. "brutacaine") as a method of restraint.
- Human family members are not allowed to be present during the euthanasia process.

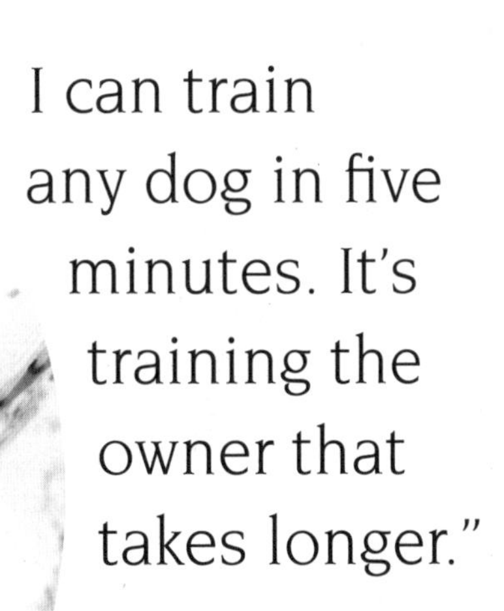

> “I can train any dog in five minutes. It’s training the owner that takes longer.”
>
> —BARBARA WOODHOUSE, AUTHOR OF *NO BAD DOGS*

4

The Office Visit: What You Can Do to Help Your Dog… and Your Vet

Okay, dear readers, now that you have found your veterinarian and clinic, here are some guidelines to govern your behavior every time you visit. This information is offered as a benefit to you, your dog, and the entire veterinary staff. I should add, though, that not every veterinarian will be keen on my advice. Doctors who schedule office visits every 10 minutes and don't care for answering questions might prefer their clients never catch wind of this book!

What to Expect During a Routine Visit

"A doctor who cannot take a good history and a patient who cannot give one are in danger of giving and receiving bad treatment."

Whether it's an annual physical exam or an appointment because of illness, you can only be helped by knowing the typical routine at your vet's office.

Your Dog's History: What Your Vet Needs to Know

Most vet visits begin with "history" taking. Your dog's medical history is comprised of two parts. First, there is his prior medical record, such as vaccinations, past surgical and medical procedures, and previous diagnoses. When you've taken your dog to the same vet clinic all along, this will already be an established part of his file, but should you change vets or be referred elsewhere, it is important to provide such background information upon the first visit.

The second part of the history consists of recent observations of your dog's health and behavior, including any abnormalities or symptoms. This is essential information to provide *every time* you visit the vet—it helps her give the "healthy dog" stamp of approval, determine if there's an issue at hand, or solve "medical mysteries" when there's no doubt something *is* troubling your pup.

The skilled vet knows just what questions to ask to elicit valuable information from her client. Your job is to provide accurate and relevant facts and to avoid getting off track with interesting but superfluous storytelling.

Getting an "A+" in History

In the late 1960s a character by the name of Sergeant Joe Friday starred in a television series called "Dragnet." He was a cop notorious for stating, "Just the facts, ma'am," when interviewing crime witnesses intent on providing irrelevant, distracting information. I often fantasize about using this line when attempting to extract a patient's history from clients who "have a lot to share." Now don't get me wrong, what such clients have to say is often interesting and amusing, but when I'm running 40 minutes behind schedule, I wish Sergeant Joe Friday were standing beside me!

Sometimes, when I meet a patient and client for the first time, I receive a long-winded dissertation about the dog's life, beginning with the day my patient was delivered from the womb. There was the broken toenail in 2004, and the time in 2006 when he vomited all night long after eating a rabbit carcass. And of course, he got into all that

SECRET FOR SUCCESS

See Your Vet Once a Year—No Matter What!

Vaccination guidelines for dogs have recently changed dramatically; most notably, the current recommendation is to vaccinate every three years rather than annually for some vaccines (see chapter 6). That's great news in terms of avoiding the unnecessary risk and expense associated with over-vaccinating. Ironically, this beneficial change may also have some negative health repercussions.

You see, we've done too good a job with those "vaccine reminder postcards" and have inadvertently programmed clients to believe that vaccinations are the most, if not the only, important part of their dog's regular visits to the vet clinic. Nothing could be further from the truth. As you might imagine, veterinarians are worried about losing the opportunity for early disease detection and treatment for their patients. It's a no-brainer that the earlier cancer is detected, the better the outcome. The same is true for heart disease, kidney disease, periodontal disease, and a myriad of other medical issues that might be detected during a routine physical examination. The annual office visit also provides time to discuss nutrition, behavioral issues, parasite control, and anything else that warrants veterinary advice.

I strongly encourage you to commit to an annual vet visit for your dog. Besides, won't it be nice for your dog to see the vet and receive lots of attention and treats—without the shots!

chocolate two years ago at the neighbors' house when he was supposed to be playing with their dog. Oh, and by the way, he sleeps on the bed between my new client and her husband every night, and it's having negative repercussions on the marital relationship. Believe it or not, I sometimes hear all this and more before I've even had a chance to say "Hello," or introduce myself!

An accurate patient history is of monumental importance when it comes to providing good medical care for your dog.

An accurate patient history is of *monumental* importance when it comes to providing good medical care for your dog. A solid history can make the difference between having to run one test, or five, to pinpoint the diagnosis. Inadequate or inaccurate information can lead the veterinarian entirely in the wrong direction. Sometimes, the diagnosis can be made on the

SECRET FOR SUCCESS

Providing Your Dog's Medical History

Do you want to provide the most helpful historical information possible in order to facilitate the best possible medical care for your pup? Do you want your vet to kiss the ground you walk on, and nominate you for sainthood? Here are the steps to make it happen:

❶ Let your veterinarian talk first and ask all her questions before you provide her with additional information you think she might need.

❷ Make sure your responses do indeed answer your vet's questions, as succinctly as possible. (Be prepared by consulting the list of symptoms and related questions beginning on p. 259 before you go.)

❸ Now is the time to be sure your main concerns about your dog are being addressed, and also your chance to offer forth additional information you believe is important and provide your "sixth sense" notion of what might be troubling him.

basis of a complete history alone. And, the quicker the diagnosis, the better for the client's pup, pocketbook, and peace of mind.

Veterinarians like to obtain their patient's history by asking you, the client, very directed and specific questions about your impressions and observations. What may sound like random irrelevant questions are, in fact, a way of obtaining the information that plays a huge role in directing what your vet does next. Veterinarians really appreciate it when the answers to our questions are concise and to the point! Sounds rather simple, doesn't it? If only it were! The examples in the sidebar on p. 63 illustrate some of the scenarios all too familiar to veterinarians—one client takes a mighty long time getting to the point, and several can't manage to actually answer the question asked (a phenomenon I fondly refer to as "answer avoidance"). You can see how we might get frustrated!

QUICK REFERENCE

Answering Your Vet's Questions: Dos and Don'ts

Just the Right Amount vs. Too Much Information

Dr. Kay: What has you concerned about Murphy today?

Ms. Succinct: "He hasn't been willing to walk up the stairs for the past week."

Ms. Talkative: "My office is upstairs, and that's where I work on my thesis. I'm getting my doctorate in adolescent psychology. My husband spends most of the evening downstairs watching television while I'm working. He usually eats popcorn while he watches TV, so Murphy likes to hang out with him. I know that Murphy's not supposed to have popcorn, but my husband only feeds him a couple of kernels, and it's not salted or buttered, so I figure that it's okay. My husband is overweight and has high blood pressure, and his doctor has told him he can't have salt and butter. Well, I typically work on my thesis until eleven o'clock or so, and my husband turns off the television at ten unless it's Thursday night. His very favorite show is on at ten on Thursday nights. So, as soon as the television goes off and the popcorn is gone, Murphy usually comes upstairs to be with me. Well, for the past week, Murphy hasn't wanted to come upstairs."

Answer Avoidance

Dr. Kay: "Have you been filling Lucy's water bowl any more or less than usual?"

Mr. On Target: "No, there has been no change."

Mr. Lost-in-Space: "Lucy has always been a big drinker. We usually go to the river three times a week, and when she's there, she spends most of her time playing in the water. She adores water."

Mr. Off Track: "We keep lots of water out for Lucy. She has a bowl in our bedroom, one in the kitchen, and one in the bathroom."

Mr. Wanderer: "Lucy has allergies, so we give her only bottled water."

Time and time again, a client's "sixth sense" about her dog has sent me in a diagnostic direction I might not have otherwise considered.

In Appendix I, p. 259, I have provided an alphabetical list of commonly observed symptoms, and the questions your veterinarian is likely to ask about them. So, when you observe one of the symptoms included, refer to these questions, and try to have your answers ready before you arrive at the vet clinic.

Once your veterinarian has had the chance to ask all of her questions, you can share your notions or "gut feelings" about what is wrong with your dog. You may not have years of veterinary training, but you know your dog much better than anyone else in the room. Time and time again, a client's "sixth sense" about her dog has sent me in a diagnostic direction I might not have otherwise considered.

What's Involved in a Physical Exam

Once you have provided your dog's history, the veterinarian will begin the physical examination. This includes:

- Assessing overall alertness and appearance
- Measuring body weight, temperature, and heart and respiratory rates
- Checking the eyes, ears, nose, and throat
- Palpating the lymph nodes
- Listening to the heart and lungs with a stethoscope
- Feeling the abdomen for lumps, bumps, or organ enlargement
- Evaluating the skin and haircoat

Every veterinarian has her own system—the actual order doesn't matter, only that, during the course of the exam, everything is evaluated. Even when you bring your dog in for a specific problem—i.e., he is limping or has an ear infection—unless he has recently had a thorough exam, you should be prepared for the whole shebang. Note: do your best to keep your dog still and quiet during the examination and refrain from talking to your vet when she is listening with her stethoscope.

> QUICK REFERENCE
> **Components of an Office Visit**
>
> 1. History: description of current health, behavior, and symptoms (if there is a problem). Note: when working with a new vet, include a summary of all prior medical issues and copies of medical records.
> 2. Thorough physical examination
> 3. Explanation of abnormal findings, if any, and discussion of options
> 4. Questions and answers

After the examination there is time to discuss any concerns. If you are in for a routine checkup and your dog is completely healthy, hurray! If a problem is what took you there, the vet will talk to you about the diagnosis (if known) or diagnostic options. She will also discuss potential treatment. Time should be allowed for you to ask questions.

Whew, that's a heck of a lot that needs to happen in a 10- to 20-minute visit! It's one of the reasons it pays to show up for your appointment on time, even a little bit early in case your vet happens to be running ahead of schedule.

The 10 Commandments of Veterinary Office Visits

Here are 10 tried and true secrets to making every visit to your dog's veterinarian exceptional for you and the entire office staff. They also directly benefit your dog's health—and nothing is more important than that.

I: *Thou shalt push thy veterinarian off her pedestal*

Much to my supervisor's chagrin, I adamantly refuse to wear a white lab coat. I agree that it would keep my clothing clean and help me stand out as a doctor, but I shun it because I believe it hinders relaxed, open conversation with my clients (I don't think dogs are crazy about white coats either). I'm referring to what is known as the "white coat intimidation factor," a phenomenon that gives the doctor an air of authority and superiority. When she is on such a "pedestal," two-way communication flounders. Medical advocacy requires active client participation, and a client who is intimidated does not feel comfortable voicing an opinion.

In most cases, the pedestal on which a veterinarian resides is a figment of the client's imagination. I'm delighted that the profession is viewed favorably but vets truly don't deserve any extra helpings of adulation. So, before you arrive at the veterinary clinic, prepare yourself to "push" the vet off her pedestal. Remember, this is a simple mind-over-matter endeavor. And, if your vet clings fast to her pedestal, consider choosing a different teammate!

II: *Thou shalt be present*

A face-to-face conversation with your vet is invariably more valuable than connecting later via phone or email. Actually being there allows you to view X-rays and see how to administer medication. And, don't forget, given the choice, your dog would absolutely, positively want you to be by his side! So, do not ask your mother, your brother, your housekeeper, the kid next door, or anyone else to pinch-hit for you. Unless you've had recent discussion with your veterinarian to arrange a procedure, if at all possible, avoid simply dropping your dog off at the veterinary hospital in the morning before you go to work or school. If this is truly necessary, consider arranging a discharge appointment during which time you and your veterinarian can talk about your dog face-to-face.

When a dog is experiencing significant symptoms or is sick, it helps to have all the decision-makers present at the time of the office visit. If this is difficult to arrange, the person present should take notes, and even consider audio-recording the conversation with the vet. This is useful since details inevitably get lost in translation—especially when traveling from spouse to spouse! Consider bringing the kids along unless they will create a significant distraction as they can be wonderfully uninhibited sources of information and keen observers of their dog's habits.

Lastly, turn your cell phone off before entering the exam room. A client who answers a call while I am discussing her dog's health isn't truly "there" with me.

III: *Thou shalt let the staff know if thy dog is aggressive*

All dogs are capable of unpredictable behavior. A savvy veterinary staff can usually peg an aggressive pooch within seconds of meeting him. Occasionally, one surprises us and bites—either a member of the staff or the client. Everyone feels terrible but it's made far worse when we learn that the client knew it could happen, but failed to warn us.

I clearly recall a nasty bite to my hand with no warning glare or growl to clue me in. As I stood by the sink washing my wound and muttering

SECRET FOR SUCCESS

Make That Dog Behave!

There are a number of ways a dog's naughty behavior can contribute to health problems. Behavioral issues are the number one reason dogs don't "work out." This is why many purebred dogs end up with adoption agencies in the first place and why adopted dogs often wind up back in shelters. Sadly, the outcome for such dogs is often euthanasia.

Untrained dogs are far more likely to dart out in front of a car, pick a fight at the dog park, or take a load of buckshot for getting into the neighbor's chicken coop! Plus, ask any vet clinic staff member, and she will no doubt agree it can be difficult to provide good quality care to a dog that lacks basic obedience training. Imagine trying to perform routine procedures such as looking in the ear canals with an otoscope, trimming toenails, or taking the temperature on a dog who is bouncing around the exam room like Tigger on amphetamines. Good manners are a must for every dog, not only for the above reasons, but for your own sanity, too.

under my breath, the client had the audacity to inform me that the same thing had happened to the last veterinarian they had seen! I momentarily fantasized about biting *her*, but showed tremendous restraint.

If your pup has previously growled or attempted to bite in a clinic setting, it is vital that you divulge this information. Trust me, withholding such knowledge is the quickest, most effective way to alienate yourself from an entire staff, and you will not be welcomed back. The flip side of this coin is that veterinarians have nothing but respect for the client who brings along a muzzle that's just the right fit.

A dog acts out of character in a hospital setting for a number of reasons. Pain, fear, a bad experience, or the need to protect his human can all provoke aggression. Fortunately, there are many humane ways to work effectively with an aggressive dog: chemical sedation or muzzling are reasonable options. Sometimes, simply separating a dog from his human subdues this aggressive tendency. Restraining with brute force (a.k.a. "brutacaine") is never warranted.

IV: *Thou shalt provide information*

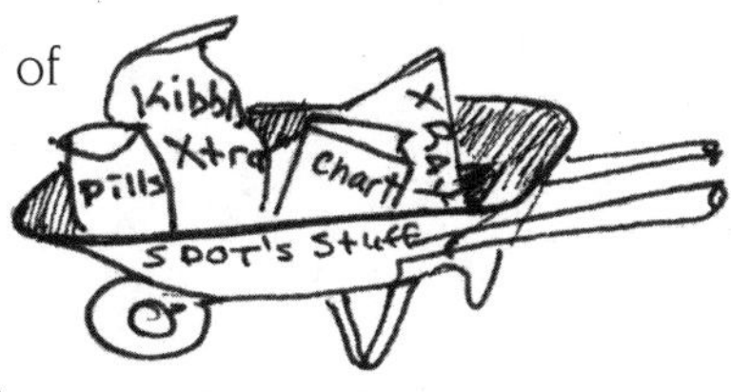

As discussed earlier in this chapter, the "history" of your dog's health, past and present, is exceedingly important, more so than many people realize. This often provides more clues for a correct diagnosis than the actual physical examination. Your vet will want to know if you've seen any changes in behavior, appetite, thirst, or energy. Report any vomiting, diarrhea, coughing, sneezing, decrease in stamina, or change in bladder or bowel habits. Reread pp. 59–64 where I explain your dog's history in detail, then do some sleuthing on the home front.

Medication and Diet

Bring your dog's current medication to every visit, so drugs and dosages can be confirmed. Your veterinarian will want the name and strength of the drug, not just a visual description of the tablet. (*Many* medications come in the form of small, round, blue pills!) All too frequently we come across a prescription that has been dispensed, or is being administered, incorrectly.

And, know the brand name of the food your pup eats. The color of the bag and name of the store where it was purchased simply won't give your veterinarian adequate information.

Prior Medical Conditions

First time visitors to a vet clinic should have in hand their dog's vaccination history as well as any medical records, laboratory test results, and X-rays that pertain to prior problems. If your dog's recent symptoms or medical history are somewhat complex, it helps to see a concise written summary of events. For example, when your dog has had a seizure disorder for the past nine months, providing a journal of the dates and duration of the seizures might be extremely helpful. By the same token, it is possible to provide *too much* information. I once received an inch-thick diary of many months' worth of a patient's bowel movements—including weights and lengths (I couldn't possibly make this stuff up).

V: *Thou shalt confess everything*

If your dog has trained you to feed him nothing but table food; if you have been sharing your own prescription medication with your pooch; if he fell out of the back of a pickup truck because he was not properly tethered; even if he has just eaten a plate of marijuana-laden brownies, you *must* convince yourself to rise above any embarrassment or awkwardness and be truthful with your veterinarian.

I once had to confess to a large animal vet that I'd fed rhododendron trimmings to my goats. Rhododendrons are toxic to goats, causing terrible abdominal distress—something *every* veterinarian learns in school, but I'd somehow managed to forget. Ingestion requires immediate and specific therapy, so my confession facilitated my goats' complete recovery, thank goodness. I still feel a wee bit embarrassed when I cross paths with the vet who saved them. Ah, the things that keep us humble!

VI: *Thou shalt pause for confusion*

It is just about impossible to do a reasonable advocacy job if you don't understand what your vet says. As the saying goes, "What we don't understand, we can make mean anything."

Most veterinarians, myself included, lapse into "medical speak" because we are so used to these terms running around in our heads. We might say to a client, "Easy is in renal failure and needs aggressive diuresis," instead of, "Easy's kidneys aren't functioning properly, and we can help him by giving him intravenous fluids." We need you to stop us in our tracks when we confuse you. If you are a "visual learner," ask your vet to draw a picture or show you what she is talking about on your dog's X-rays, lab reports, or ultrasound images. Remember, *always* "pause for confusion"—when you don't understand, stop and get clarification.

SECRET FOR SUCCESS

"Detox" Your Dog's Environment

The average home and yard contain plants, chemicals, food products, and household substances that are potentially poisonous for dogs. You've probably heard about the dangers of wild mushrooms, but did you know that iris, azalea, and foxglove are poisonous plants? A single sip of antifreeze can be deadly (it has a sweet taste that many dogs love). Rat and mouse bait's toxicity isn't limited to rodents. Chocolate can cause neurological symptoms, and raisins, kidney failure. The way to "puppy proof" is to research which plants, foods, chemicals, medications, and household products could be hazardous. A great place to start is with the ASPCA Animal Poison Control Center (www.aspca.org) and see my list on p. 374.

It's ideal to get rid of all potential toxins from your dog's environment, but for those that cannot be removed (I'd never ask you to deprive yourself of chocolate), take precautions to completely eliminate your pup's access to them. This is especially true if you happen to have a dog that likes to "explore" everything in his path or is intent on eating almost anything that doesn't eat him first!

VII: T*hou shalt share thy concerns*

Most veterinarians do what they do because they appreciate how much dogs mean to their humans. Who better, then, to empathize with you? To help you, your vet needs you to tell her your particular worries and concerns:

❶ Are you feeling scared or angry? (Anger is a normal stage of the grief process—many people experience it in response to a dog's illness.)

❷ **Are financial limitations creating a roadblock?**

❸ **Are you convinced your dog has a terminal disease?**

❹ **Are you terrified by the thought of anesthetizing your dog because a beloved pet once died unexpectedly while under anesthesia?**

❺ **Are you receiving pressure from family members or co-workers to put your dog to sleep, but you don't think it's time yet?**

Your vet will be better able to understand your reasoning if she knows how you are feeling, and you will receive a much-needed dose of empathy.

Financial Matters

It's never easy discussing financial worries—candor suffers because the subject is often awkward and much too personal. Clients feel guilty and worry about being judged when cost needs to be a factor in medical decisions. Be aware, though, you should discuss this matter up front. Be sure to get an estimate before services are provided so as to avoid any unpleasant surprises. Ask about payment plans or prioritization of services. Most veterinarians are willing and able to provide reasonable financial options. For more on this subject, see p. 225.

VIII: T*hou shalt ask questions*

Asking questions is the most resourceful way to be your dog's medical advocate. In the heat of the moment, when you have just received some disconcerting news, a child is tugging at your arm, and your dog has just lifted his leg rather too near the veterinarian, it is easy to forget the important questions you were meaning to ask. It pays to write

them down beforehand. No doubt, you will do some homework and research (including reading this book) when you get home, and you will invariably think of more questions you should have asked. No problem. Veterinarians expect clients to call with questions after they've had some time to process and ponder the information they've received.

I address the question-asking issue directly in chapter 7, p. 131, and Appendix II, p. 283. There, you'll find all sorts of things to ask your vet, depending on the information you've been given. I recommend you bring this book with you to your next office visit so it is conveniently at hand as a reference.

IX: *Thou shalt treat the entire staff well*

I get really peeved when I learn that a client, who has been sweet as can be with me, has been abrupt, condescending, or rude to one of my staff. Everyone deserves to be treated with equal respect, and, without a doubt, the entire staff will know if this has not been the case! Likewise, a client who has been respectful and gracious will have the "red carpet" rolled out the next time she visits.

X: *Thou shalt always come away with a plan*

What do I mean by this? It is this simple: every time you talk with your veterinarian, be sure you know exactly when and how you will next communicate. Consider the following examples:

❶ Your six-year-old Norwegian Elkhound has just had his annual checkup, and, much to your delight, everything is completely normal. The "plan" is to bring him back in one year for his next "annual."

❷ Your three-year-old Chihuahua-Jack Russell Terrier mix has just been evaluated for coughing, and prescribed an antibiotic and cough suppressant. The "plan" is to call the hospital in one week to report whether or not the cough has fully resolved. If not, chest X-rays and a blood test will be scheduled.

❸ Your Golden Retriever puppy has a heart murmur. Ultrasound reveals a problem with the mitral valve in his heart. Future prognosis is uncertain. The "plan" is to repeat the ultrasound in six months, or sooner if coughing or decreased stamina is observed.

❹ Your terrier mutt just had surgery to remove bladder stones. At the time he is discharged from the hospital, the "plan" is to feed him a special diet to prevent stone reformation, return in two weeks for removal of the stitches, and schedule a two-month follow-up to recheck a urine sample.

Vets often fail to provide clear follow-up recommendations and well-intentioned clients often fail to comply with them. Do your best to solidify the "plan" and put it in writing. You'll be glad you did.

"Every great advance in science has issued from a new audacity of the imagination."

—JOHN DEWEY,
AMERICAN PHILOSOPHER,
PSYCHOLOGIST, AND
EDUCATIONAL REFORMER

5

Wow, I Didn't Know You Could Do That on Dogs!

Choosing the title of this chapter was a simple task, because I'm so used to hearing, "Wow, I didn't know you could do that on dogs!" Everyone knows that veterinarians vaccinate, neuter, treat infection, and mend broken bones. What they may not realize is that we get to work with all kinds of high-tech diagnostic toys and gadgets (oops—I mean medical instruments and equipment!)

Veterinary medicine is never more than a few steps behind its human counterpart—and it often paves the way for advances in the human field.

I get a chuckle out of knowing that medical doctors are often the most astonished at our advanced capabilities. They're surprised to learn we are doing many of the same things with dogs that can be done with their human patients. What these doctors may not realize is that veterinary medicine is never more than a few steps behind its human counterpart. In fact, it often paves the way for technological advances in the human field. For example, stem cell therapy is currently being used to treat some canine diseases. Meanwhile, the clinical application of human stem cell therapy has been bogged down in a quagmire of religious and bioethical controversy.

On the pages that follow are descriptions of some amazing diagnostic and therapeutic technologies available for use on dogs. Some

BREEZY It was love at first sight when Edward laid eyes on his Breezy girl for the first time. She couldn't have been more than seven weeks old when he spotted her wandering around a convenience store parking lot with neither human nor canine caretaker in sight. The overly confident, black-and-tan Australian Shepherd-something-or-other had no idea that her life was in danger as she navigated her roly-poly body between cars, saying hello to every person she saw.

Edward brought Breezy home and tenderly loved and cared for her for the next 12 years. During her twelfth year, Breezy developed a cough, one that worsened over the course of a few weeks. She was examined by her veterinarian who took chest X-rays and discovered a lung tumor. Surgery to remove the mass was recommended with the hope that it would provide long-term remission (a cancer-free state) if not a cure. Based on Breezy's overall good health, Edward opted to proceed, and his vet referred him to my hospital for the surgery.

The Breezy I met was a huge, gregarious and affectionate beast—the kind of dog that could make even the grumpiest face crack a smile. She appeared robust and strong—an ideal candidate for surgery. I fully supported the notion of surgery but encouraged Edward to first allow one more test: a computed tomography scan (CT or CAT scan) of Breezy's chest cavity to be certain that the X-rays were telling us the whole truth. (When it comes to lung masses, X-rays sometimes underestimate their number. CT scans are far more sensitive at picking up smaller tumors.) I remember Edward's response very clearly. He said, "Wow, I didn't know you could do CAT scans on dogs!" He was game to give it a go.

Much to our chagrin, we discovered that Breezy was harboring not just one, but three tumors, all located in different lung lobes. These terribly disappointing results dashed any hopes we had for surgically treating her cancer. While it's feasible to take out one or two lung tumors, removal of three, in different lobes, simply wasn't a reasonable option. Without significant optimism that chemotherapy would be of benefit either, I sent Breezy home on a course of palliative (hospice-like) therapy, which consisted of a cough suppressant and pain medication when needed.

Edward catered to Breezy's progressively finicky appetite until the time she was euthanized approximately three months later. Although he was heartbroken, he derived profound comfort from the knowledge that he had not subjected his beloved Breezy to a needless major surgical procedure (which would have proved unsuccessful) during the last few months of her life. At the time, I remember thinking to myself, "Thank goodness we can do CAT scans on dogs."

are brand new while others have been around for a long time, continuously transforming to keep pace with technological advances (compare the television set you had in the 1980s to the one you watch today).

21% of dog owners would travel 1,000 miles or more to obtain specialty care for their pet.

Veterinary Specialists

What Are They?

Did you know that you can take your dog to see a "specialist," such as a cardiologist, neurologist, ophthalmologist, internist, dermatologist, or oncologist? A veterinarian cannot become a specialist simply because she has an interest or aptitude in a particular field of medicine. The term is reserved only for those with the desire and fortitude to have continued their education beyond four years of veterinary school—they must also complete a minimum three-year internship and residency training program, author publications, and then pass some insanely rigorous exams. The culmination of this extended, specialized training and testing is typically referred to as "board certification" or "certification" in the chosen field of study. You can recognize an individual's qualifications by the initials or words that follow her name (see more about this on p. 215).

Vets who choose to specialize tend to have a strong drive to become highly adept at their particular field of medical interest.

Where Are They?

The world of veterinary specialists has grown by leaps and bounds. Just like Starbucks®, if there's not already a group of specialists in your neighborhood, there likely will be soon! Veterinary specialists are found in university teaching hospitals and in some private practices. They often "cohabitate," sharing specialty staffing, equipment, and laboratory services with specialists in different fields. When this is the case, you, the lucky client, end up with access to multiple specialists under one roof. Not only is this convenient, it also focuses a lot of brainpower and experience on your pup—group discussions about patients (medical rounds) typically occur once or twice daily.

When Do I Need Them?

Most often people take their dog to a specialist following referral from their family veterinarian, similar to the way things work in human medicine. It is likely your vet has established relationships with local specialists—the kind she would trust to take good care of her own dog. I like to think of the primary care veterinarian as the "general contractor," and specialists as her "subcontractors." Ideally, the family vet remains an integral part of the dog's health-care team even when the specialist has temporarily taken over. Most specialists are conscientious about keeping primary care referring veterinarians informed and involved.

Ideally, the family vet remains an integral part of the dog's health-care team even when the specialist has temporarily taken over at the helm.

You do not need to be referred by a veterinarian in order to see a specialist. Self-referrals—setting up appointments entirely on your own—are perfectly okay! And, depending on where you live and the number of specialists available, you will likely be able to see one within a matter of days or weeks. If you are concerned that your dog's condition is urgent, always make that known when setting up the appointment.

If you need help finding a specialist in your area, consult the sidebar in chapter 9, p. 218.

QUICK REFERENCE

An Advanced Directive for Your Dog

You've been conscientious—all the official paperwork is in place so that, in the event of your death, it will be undeniably clear who will inherit your possessions and assume guardianship of your children. What about your dog? What will happen to him? If you haven't prepared an advanced directive for your precious pup, here are some tips for getting started:

❶ Select the person you want to assume guardianship (this might be more than one person if more than one dog is involved). Confirm their willingness to take on this responsibility—be specific about your wishes for his quality of care and your philosophy about medical treatment and euthanasia.

❷. Prepare all of the official paperwork just as you would for other advanced directives.

❸ Set up a trust fund to care for your dog's future needs. Providing for the guardian will allow the guardian to provide for your dog.

Ultrasound

What Is It?

Ultrasound imaging is one of my favorite diagnostic tools. It provides so much information with virtually no risk to the patient—a way of "nonsurgically" looking inside the body (the same technique commonly used on pregnant women to evaluate the health of the fetus). It provides a three-dimensional, real-time image of internal organs rather than the two-dimensional still shots from an X-ray. For example, an X-ray shows the silhouette of the heart, but ultrasound takes us right inside each of the four chambers where we can evaluate size, how well the heart muscle is contracting (beating), and the function of the heart valves. It's truly an amazing thing to watch!

Ultrasound provides a way of "nonsurgically" looking inside the body.

Ultrasound is performed by using a hand-held probe that produces sound waves, which travel into the body, deflect off internal structures, then bounce back to the transducer. The sound waves bouncing back to the probe create the image we see on the screen. Each tissue type deflects the sound waves differently, thus allowing abnormalities to be detected. For example, a mass in the spleen, an abscess in the liver, or a stone in the bladder produce obvious changes to the normal images of those body parts.

When Is It Used?

Not only does ultrasound locate abnormalities in internal tissue, it also provides the opportunity to take measurements. When performing an evaluation of the heart (*echocardiogram*), the ultrasonographer can measure the heart muscle's ability to contract, blood flow across the heart valves, and the heart chamber's size. This information, when abnormal, helps determine the prognosis, and the procedures and medication that will provide the most benefit to the patient.

Ultrasound is also used to help collect tissue samples; for example, when I biopsy an enlarged lymph node within the abdomen, I create a tiny incision in the skin (so small that no stitches are required) through which the biopsy instrument is passed. I then use ultrasound to help me visually guide the instrument into the lymph node.

Ultrasound studies are most commonly performed on the heart, abdominal cavity, and smaller body structures—such as the eyes, thyroid glands, blood vessels, and testicles. And, of course, ultrasound is used for pregnancy detection, as well as to assess the health of the fetuses as the dog's pregnancy progresses.

Where Is It Available?

More and more family veterinarians—and most specialists—are

incorporating ultrasound technology into their practices. The vet performing your dog's study should ideally have a lot of experience as ultrasound proficiency has a monumentally steep learning curve.

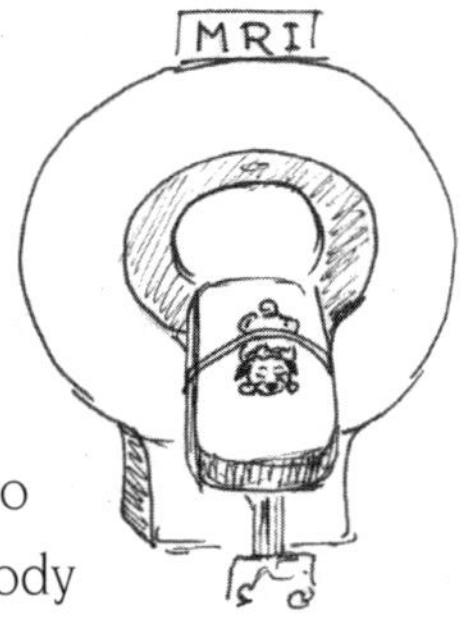

CT and MRI Scans

What Are They?

CT or CAT (computed tomography) and MRI (magnetic resonance imaging) scans create highly detailed images (more so than ultrasound or plain X-rays can provide) of any and all body parts. Whereas an X-ray shows a two-dimensional silhouette of the body part in question, CT and MRI machines use computer technology to capture a series of images (like slicing a loaf of bread) to create exquisite three-dimensional images. For example, an X-ray of a dog's spine looks very much like the actual skeleton. CT and MRI scans allow visualization of the spinal canal (the bony tunnel that houses the spinal cord), the spinal cord itself, and the disks between the vertebrae.

The CT scanner utilizes X-ray beams that pass through the body to a detector device feeding information into a computer. The computer analyzes this information based on tissue densities, then conveys this data to a monitor for all to see. An MRI machine obtains its image by using gigantic magnets. Within the body there are millions of negatively and positively charged particles. When they are exposed to the electromagnetic waves produced by the MRI equipment, these particles act like mini-magnets. The MRI computer is capable of interpreting changes in magnetic forces and transforming them into a highly detailed three-dimensional anatomical image. Information provided by an MRI scan is, generally, a bit more sophisticated and detailed than that from CT scanning.

Contrast agents are commonly administered during scans to enhance visualization of various structures. In short, MRI and CT scans provide information that, in the past, could only be obtained by surgical exploration of the body part in question.

When Are They Used?

These scans are commonly used to evaluate the brain, spinal cord, joints, abdomen, and chest cavity. They are also useful for diagnosis of tumors and traumatic injuries. CT scans are commonly used to identify cancerous, infectious, or inflammatory diseases within the limbs, abdominal, and chest cavities. MRI is commonly used to evaluate the central nervous system—the brain and spinal cord.

Where Are They Available?

They are commonly found in university teaching hospitals. More and more private specialty veterinary centers are making room under their roofs and in their budgets for CT and MRI scanners. It goes without saying that CT and MRI scans are more expensive than plain X-rays, primarily because they involve large, complex machinery with expensive maintenance requirements.

Digital Radiography

What Is It?

We've come a long way from the old hand tanks where we had to sit in a tiny dark room for five to 10 minutes at a time, inhaling chemical vapors while developing a set of X-rays. Digital radiography is rapidly catching on in the worlds of both human and veterinary medicine. Instead of using traditional photographic film, digital X-ray sensors are used. The beauty of digital radiography is that one can manipulate the pixel shades to correct density and contrast within the image. Thus, a less than ideal exposure can be transformed into a quality end product.

The beauty of digital radiography is that you can manipulate the pixel shades to correct density and contrast within the image.

Not only does digital imaging improve the diagnostic capabilities of X-rays, it also cuts down on the numbers of "reshoots," thereby decreasing the patient's exposure to radiation. By virtue of the fact that digital images are stored in computer memory, they can be readily retrieved for viewing and transmitted electronically between doctors in different locations. And, as if all this wasn't enough, digital radiography is good for the environment—no more eco-unfriendly X-ray film and developing chemicals to be disposed.

When Is It Used?

Digital X-rays can be used any time film X-rays would normally be taken. The abdomen, chest, spine, and legs are the body parts most commonly imaged.

Where Is It Available?

Most specialty and many general vet practices have made the switch to digital imaging.

44% of dog owners would spend $3,000 or more on their dog's medical care.

Nuclear Medicine

What Is It?

Nuclear medicine is yet another type of imaging that helps figure

out the internal workings of the body. An injection of a radioactive material is administered, and then a picture of the dog is taken using a specialized piece of equipment called a *gamma camera*. The type of radioactive material used dictates which body parts will "glow." For example, with bone scans, cancerous lesions in the bone are highlighted. When an image of the thyroid gland is desired, a radioactive material that will concentrate in the thyroid gland is used.

When Is It Used?

In addition to the types of scans mentioned above, nuclear medicine studies are ideal for diagnosing a blood vessel abnormality in dogs called a *portosystemic shunt*.

Where Is It Available?

Radioactive materials can be handled only in licensed facilities (university teaching hospitals and some specialty private practices) and there are a lot of hoops to jump through in order to get that license! People can go home immediately following participation in a nuclear medicine study, however, veterinary rules and regulations stipulate that dogs that have received radioactive pharmaceuticals must remain isolated in the hospital for a minimum of 24 hours. Why the difference? I am certain that someone, somewhere, involved in government regulations knows the answer to this question!

Fluoroscopy

What Is It?

Fluoroscopy is an X-ray that appears as a movie rather than a still shot. It is particularly helpful when it's important to evaluate the motility, or movement, of a particular body part. For example, the esophagus is the muscular tube that transports food and water from the mouth down into the stomach. It accomplishes this with a set pattern of rhythmic muscular contractions called *peristaltic waves* (kind of like squeezing sausage out of its casing). If a dog is suspected of having a problem with these normal contractions, a fluoroscopic movie of the esophagus—after a bite of food is swallowed—is far more likely to be diagnostic than a still shot from plain X-ray.

When Is It Used?

Fluoroscopy is commonly used to observe the flow of contrast material—dye—around the spinal cord (*myelogram*) in dogs with spinal cord disease. It is the test of choice for diagnosis of a collapsing trachea and swallowing disorders. Veterinary cardiologists love fluoroscopy, because it is the technology that allows them to accurately place pacemakers, catheters, coils, and pressure monitoring devices within the heart and its surrounding blood vessels.

Where Is It Available?

If your dog could benefit from a fluoroscopic study, you will likely be referred to a veterinary teaching hospital or private specialty center.

QUICK REFERENCE

The DNR Directive

If your dog is hospitalized and in critical condition, the veterinarian may ask how you feel about attempts to resuscitate your dog should he cease breathing or should his heart stop beating. Ask her to explain what's involved in resuscitation and the likelihood of it being successful. Also consider the overall prognosis for your dog's illness.

If you decide to forego resuscitative measures, your pup's medical orders will be highlighted with the "DNR" (do not resuscitate) directive and he will be allowed to pass away naturally, if and when the time comes.

Interventional Cardiac Procedures

What Are They?

You probably thought that pacemakers were only for people. Not anymore! Just like humans, some dogs have heart rhythm abnormalities that can only be corrected with the use of a pacemaker. The pacemaker "paces" the heart, causing it to beat within a particular range.

Placement of cardiac pacemakers is one of a handful of high-tech interventional cardiac procedures. Using fluoroscopy (see p. 87), various instruments can be guided into the heart chambers and their surrounding major blood vessels for diagnostic and therapeutic purposes.

When Are They Used?

No longer is it necessary to perform open chest surgery to repair a birth defect called a *patent ductus arteriosis* (PDA). Nowadays, using interventional cardiac techniques, a nifty little coil apparatus can be positioned just so within the defect, and there you have it, the problem is corrected.

Interventional procedures are also used for repairing heart valves, pacemaker implantation, taking pressure measurements within the heart, and treating certain types of heartworm disease.

Where Are They Available?

A great deal of specialized high-tech equipment, as well as a board certified veterinary cardiologist, are necessary for such fun and games. Look for a university teaching hospital or private specialty practice.

Endoscopy

What Is It?

Endoscopy refers to the use of a long telescopic device (endoscope) to "nonsurgically" evaluate structures inside the body (any of you who has undergone a colonoscopy knows exactly what I'm talking about). The image at the end of the endoscope is transmitted to either an eyepiece or video monitor that looks very much like a television screen.

When an endoscopic procedure is performed, the scope is passed through natural orifices (nose, mouth, rectum) or entry points into the internal workings of the body. No new openings are created. Endoscopes go into all types of interesting places in a dog's body, including the throat, nose, windpipe, lungs, esophagus, stomach, small intestine, large intestine, vagina, uterus, urinary bladder, and urethra (the tube that leads from the bladder to the outside world). They come in flexible and rigid varieties. The flexible scopes are used when one must navigate curves and corners, such as those within the intestinal loops, while the rigid ones are utilized in straighter pathways, such as a dog's nasal passages.

When Is It Used?

When viewing an endoscopic procedure, it's easy to feel like a miniature adventurer traveling on an anatomical safari. Not only do you get to visualize the terrain, but all kinds of tissue samples can be collected along the way using tools that are passed through ports within the device. The beauty of endoscopy is that, in many cases, it can accomplish what would otherwise require a surgical procedure. Although the patient must be anesthetized for most types of endoscopy, the potential for complications and the time needed for full recovery are far less than with surgery.

Endoscopy is commonly used to detect inflammatory, infectious, and cancerous conditions within the respiratory, gastrointestinal, or urinary tracts. It is also used for artificial insemination techniques. And for those dogs with undiscriminating tastes, endoscopy can be used to retrieve foreign objects from the esophagus or stomach. I've removed some memorable foreign bodies throughout the years, ranging from expensive jewelry (worn again after retrieval), to fishhooks, to pantyhose. I clearly recall the horror of endoscopically peering into the stomach of a very large, aggressive dog (anesthetized, thank goodness), and seeing what I thought was a human hand, ear, and hair. It turned out to be the chewed remains of a troll doll!

The beauty of endoscopy is that, in many cases, it can accomplish what would otherwise require a surgical procedure.

Where Is It Available?

Some general practices and most specialty veterinary centers and university teaching hospitals provide endoscopic services.

Laparoscopy, Thoracoscopy, and Arthroscopy

What Are They?

These procedures very much resemble endoscopy (see p. 89) in that a telescopic device is used to look inside the body. Here's the difference: With regular endoscopy, the scope is introduced into the body via a natural orifice. When performing laparoscopy, thoracoscopy, or arthroscopy, a rigid scope is introduced through a small incision into the abdomen, chest cavity, or joints, respectively. These "oscopies" provide a fabulous new way to look around inside the body. Although they are considered "surgical" procedures, the degree of "invasiveness" and recovery time is far less than with conventional surgery.

When performing laparoscopy, thoracoscopy, or arthroscopy, a rigid scope is introduced through a small incision into the abdomen, chest cavity, or joints, respectively.

When Are They Used?

Laparoscopy is used to visualize organs within the abdominal cavity (liver, gall bladder, stomach, intestines, pancreas, spleen, kidneys, and urinary bladder) and obtain biopsy samples. In the years to come, laparoscopy may replace more traditional surgery for procedures such as removal of the gall bladder and spaying (removal of uterus and ovaries).

Thoracoscopy is used to visualize the heart and lungs within the chest cavity. Currently, it is most commonly used to perform surgery on the pericardial sac, the impermeable membrane that surrounds the heart.

Arthroscopy is used to diagnose and treat joint ailments. Just as in human medicine, arthroscopic surgery can be used in place of traditional surgery to repair a variety of joint and ligament abnormalities.

Where Are They Available?

Laparoscopy, thoracoscopy, and arthroscopy procedures are available mostly at private specialty clinics and university teaching hospitals.

Chemotherapy

What Is It?

The term "chemotherapy" refers to the use of drugs or chemicals as treatment for cancer. Why do we administer chemotherapy to dogs? There are two reasons. First of all, it works well with many types of cancers. It is often extremely successful at derailing, if not annihilating, malignant cells. Secondly, the unpleasant side effects that people experience are far less common in dogs (even though many of the same drugs are used). If this were not the case, I think you'd be hard pressed to find many veterinarians willing to administer chemotherapy to dogs. Our oath of office talks about alleviating our patient's suffering, not prolonging it.

The unpleasant side effects that people experience with chemotherapy are far less common in dogs.

When Is It Used?

Although chemotherapy doesn't work well on all types of cancer, it is highly effective against certain forms (see chapter 8, p. 188).

Where Is It Available?

Some family veterinarians feel comfortable administering specific types of chemotherapeutic drugs. Many prefer to refer their patients to a board certified veterinary oncologist or internist.

Radiation Therapy

What Is It?

In dogs—as in people—radiation therapy is an effective treatment for some types of cancer. Photons, electrons, and gamma rays are used to damage and destroy proliferating cancer cells. Radiation can be used as the sole treatment modality for some types of cancer, but more commonly it is used in conjunction with surgery or chemotherapy. At its best, radiation therapy results in a cancer cure. It can also be used to provide a temporary cancer remission, or alleviate the pain associated with certain types of cancer.

Where Is It Available?

Radiation therapy is provided at most veterinary teaching hospitals and some private specialty centers under the guidance of a board certified veterinary radiation oncologist. See chapter 8, p. 191 for more detailed information about radiation therapy.

Cancer Vaccine

What Is It?

Up until recently, oral cavity melanoma was considered one of the worst types of cancer a dog could have. (Melanomas are a cancerous proliferation of *melanocytes*—pigment producing cells in the body.) Neither surgery or chemotherapy, nor radiation therapy were successful at slowing the progression of, much less stopping, this awful disease. An oral cavity melanoma diagnosis has always spelled nothing but a terrible prognosis with a survival time rarely more than a few weeks to a few months.

However, during 2007, a vaccine was licensed for use in treating dogs with this condition and the results have been astounding! Dogs who receive the vaccine are experiencing good quality survival times in excess of a year. The vaccine is genetically programmed to convince the dog's immune system to attack the melanoma cancer cells while leaving the body's normal melanocytes alone.

The melanoma vaccine is an incredibly exciting and successful step in the realm of cancer therapy. I've no doubt we'll be seeing more cancer vaccines in the future, as they have the potential to revolutionize how we fight this disease.

The melanoma vaccine is an incredibly exciting and successful step in the realm of cancer therapy.

Where Is It Available?

At the time of publication the melanoma vaccine is available through board certified oncologists (found in some private specialty practices and university teaching hospitals).

Blood Transfusions

What Is It?

Giving blood to dogs is a daily occurrence in some busy veterinary hospitals. Blood for transfusions can be obtained from onsite donors (the price a dog pays for belonging to a hospital employee), or from a canine blood bank where donors have been thoroughly screened for blood-borne diseases—much the same way human donors are screened for HIV and hepatitis.

Transfusions can consist of whole blood (exactly as it is in the donor dog), or of a single whole blood component. For example, an anemic dog is usually in need of only red blood cells. Dogs with blood-clotting abnormalities are often transfused with plasma—the protein-rich component of blood that contains all the necessary clotting factors.

To avoid transfusion reactions, small samples of the donor and recipient blood are often cross-matched for compatibility in advance of the transfusion.

Where Is It Available?

Some family vets feel comfortable overseeing the transfusion process. Others prefer to refer their patients in need of a transfusion to a specialty or teaching hospital where blood products are often more readily available.

Amazing Surgical Procedures

What Are They?

Did you know that veterinary ophthalmologists can perform cataract surgery to enable a blind dog to once again chase a ball (and the neighbor's cat)? Did you know that back surgery on a dog paralyzed from disk disease can restore normal gait and function? And, did you know that hip replacement surgery is relatively commonplace for dogs with advanced arthritis these days?

Perhaps because I am an internist by trade and I do no surgery whatsoever, I'm utterly in awe of the tasks performed by my surgical colleagues. Board certified veterinary surgeons have attained the highest level of expertise. The specialists in my practice are constantly busy removing diseased lung and liver lobes; "replumbing" faulty urinary tracts; unraveling twisted bowel loops; repairing birth defects; removing things that don't belong (foreign bodies, bladder stones, cancerous growths); freeing-up things that do belong; and restoring normal anatomy in the aftermath of hit-by-car traumas, dog fights, lacerations, and any of the other dozens of ways dogs manage to get into trouble. Never assume that something is surgically impossible, until you hear it directly from the mouth of a board certified veterinary surgeon!

Where Are They Available?

Though more common surgeries may be par for the course at your vet clinic, most new or complex procedures are handled by board certified surgeons who typically set up shop in a veterinary teaching hospital or private specialty center.

Lithotripsy

What Is It?

Just like people, dogs occasionally develop urinary tract stones. Sometimes these stones develop secondary to chronic infection. Others are referred to as "metabolic stones" because they form as a result of excessive secretion or accumulation of various mineral compounds within the urine.

The traditional way of getting rid of such stones has been surgical removal. Nowadays, there is the lithotripsy option. Lithotripsy uses shock waves or laser technology to crush the stones into sand-like particles that can harmlessly pass out of the body in the urine.

Where Is It Available?

At the time of publication, lithotripsy was available only at some university veterinary teaching hospitals.

Pain Management

What Is It?

Pain management has become a huge focus of attention and continuing education within the veterinary profession. And thank goodness for that! I'm embarrassed to tell you that when I graduated from vet school, pain management was not used routinely following spay and neuter procedures. What were we thinking!

Fortunately, as a profession, we've awoken from our ignorant slumber, and come to recognize the importance of pain prevention. When it comes to surgery, the goal is to be proactive. This means getting pain medication "on board" before the pain has a chance to begin. We have access to a broad array of pain-management options, including nonsteroidal anti-inflammatory medications formulated specifically for dogs (human NSAIDs can be toxic to dogs' kidneys and stomach), narcotics (also available in extended-release patch form), acupuncture, chiropractic care, and rehabilitation therapy.

I'm proud of my profession for stepping up to the plate and taking pain management seriously.

Where Is It Available?

Nowadays, in order for a vet clinic to achieve American Animal Hospital Association accreditation, there must be proof that hospitalized patients are carefully assessed for pain throughout their time spent there. This is exactly as it should be, and I'm proud of my profession for stepping up to the plate and taking pain management seriously. I encourage you to include questions about pain management policies when you interview prospective vets (see chapter 3, p. 39).

Physical Rehabilitation

What Is It?

The goal of physical rehabilitation is to improve a dog's quality of life by eliminating, or at least minimizing, the physical limitations caused by injury, disease, and the aging process.

Physical rehabilitation for dogs is, simply put, the equivalent of physical therapy for humans (we can't officially refer to it as such because the "right" to the term "physical therapy" is reserved for human medicine only). The goal of physical rehabilitation is to improve a dog's quality of life by eliminating, or at least minimizing, the physical limitations caused by injury, disease, and the aging process. Rehabilitation techniques include, but are not limited to massage, stretching, and therapeutic and strengthening exercises.

Just as in human physical therapy, specialized equipment is often used. My personal favorite is the underwater treadmill. Picture a really huge aquarium tank and a happy dog paddling away, experiencing mobility and muscle strengthening that are not possible for him on dry land. The speed of the treadmill and depth of water are adjusted according to the patient's specific needs.

When Is It Used?

Which patients are good candidates for rehab? Dogs struggling with arthritis, soft tissue and muscle injury, certain neurological and orthopedic diseases, wound healing, and athletic-performance issues all benefit. It is commonly used following orthopedic and spinal cord surgery, and work on a treadmill is a beneficial part of weight loss programs for dogs too heavy to exercise on dry land. Rehabilitation therapists can now help dogs with permanent hind-end dysfunction regain mobility by fitting them with a customized cart (a doggie wheelchair).

Where Is It Available?

Specialized training is required—the provider need not be a veterinarian, but should hold certification in canine rehabilitation therapy.

Advanced Dental Procedures

What Are They?

It wasn't many years ago that dental care for dogs consisted of cleaning the healthy teeth and pulling the unhealthy ones. Not anymore! Nowadays, many high-tech dental procedures are performed, including root canal surgery, orthodontics, periodontal surgery, crowns, and fillings. Your dog now has access to pretty much all the same dental procedures you do!

Where Are They Available?

Some family veterinarians attain the continuing education necessary to develop expertise in advanced dental diagnostic and therapeutic procedures. Others prefer to refer their patients to board certified veterinary dentists.

37% of people have had their dog's teeth cleaned at a vet clinic.

Prescription Diets

What Are They?

If you think that prescriptions are only for medications, you're in for a big surprise. Vets also "prescribe" oodles of specially formulated canine diets that have been enormously successful at helping to manage a number of diseases. In most cases, the diet is prescribed for use in conjunction with other treatment. For some diseases, diet alone is the recommended therapy.

When Are They Used?

Prescription diets are available for prevention and treatment of skin allergies, gastrointestinal disorders, organ failure (heart, liver, kidneys), dental disease, bladder stones, obesity, diabetes, arthritis, cancer, and brain-aging.

Where Are They Available?

Several dog food manufacturers have jumped on the prescription diet "bandwagon," so there is a large selection. Remember to consult your veterinarian for recommendations.

Dialysis

Sometimes dialysis is the only thing capable of keeping the dog alive while the kidneys heal.

What Is It?

You've probably heard the term "dialysis" used when discussing people with kidney failure. Several times a week, the patient is connected to a machine that detoxifies the blood, in essence, doing the work of the failing kidneys. Believe it or not, dialysis is also available for dogs in kidney failure.

When Is It Used?

Unlike a human, who can be maintained on dialysis for years, for dogs it is a short-term treatment only. The goal is to maintain the patient for a period of days or weeks while his kidneys have a chance to heal from whatever disease or toxin damaged them. For example, ingestion of ethylene glycol, the substance in antifreeze, is a common cause of acute kidney failure in dogs (unfortunately, most brands of antifreeze have a sweet taste that dogs love). Sometimes dialysis is the only thing capable of keeping the dog alive while the kidneys repair.

Where Is It Available?

At the time of publication, there are only a handful of dialysis centers in the US, and all are affiliated with university teaching hospitals.

Stem Cell Therapy

What Is It?

Yes indeed, we are using stem cell therapy in veterinary medicine today! Canine stem cell, or regenerative cell therapy, is being used to hasten the repair and healing of injured tendons and ligaments as well as treat arthritis. The patient's own *adipose tissue* (fat) serves as the source of stem cells. A fat sample is collected by the veterinarian and sent off to a laboratory for the intricate process of growing and harvesting stem cells. They are returned to the vet who injects them into the injured or diseased joint, or other area of the dog's body. We are just at the starting point for some utterly amazing results obtained with canine stem cell therapy!

SECRET FOR SUCCESS

Think Before You Feed

What better way to prevent disease than by feeding your dog a balanced, complete diet? The trouble is there are so many advertised brands and special diets to choose from, sometimes it seems you could feed a different dog food every single day of your pup's life!

There are the standard, commercially prepared dry and canned-food diets made specifically for various life stages (puppy, adult, senior). Some are formulated based on the dog's job description (pregnant, nursing, performance dog, couch potato). Prescription diets are designed to be part of the treatment for a variety of diseases including: arthritis, diabetes, obesity, bladder stones, senility, and disorders involving the kidneys, heart, liver, and bowel. For treatment of food allergies, there are novel "protein diets" containing duck, venison, fish, rabbit, and yes, even kangaroo. Raw food diets aim to avoid preservatives and processing and mimic what dogs would consume in the wild. The most recent arrival on the dog food scene is the "breed diet" marketed to prevent various breed-specific maladies. And of course, some people believe "homemade" is best!

It's no wonder that choosing the best diet for your dog can be a daunting task. The good news is that for the vast majority of dogs, there is no one perfect diet, but many different excellent choices. So, it's up to you to determine how much energy you wish to invest in "diet" research. Here are some suggestions:

❶ Solicit recommendations from knowledgeable people, and educate yourself about your dog's nutrition in the same way that you would for other aspects of his health.

❷ Reputable dog food companies invest time, science, and research in order to produce a balanced, safe, and complete diet. Look for the official label indicating the food is "complete and balanced" as established by the AAFCO (Association of American Feed Control Officials) standards. The AAFCO is to dog food what the FDA is to prescription medication. Your local health food store may sell their own brand of dog food, so it's normal (and easy) to assume this pet food is good because of where it originates. My concern is this: how can a small, independently owned shop have the background, know-how, and capital available to research how to create a truly balanced dog food?

SECRET FOR SUCCESS CONT.

❸ Consider your dog's stage of development. The nutritional needs of a puppy are vastly different from those of a "senior citizen."

❹ Take into account what your dog does for a living. Is he a champion agility dog, or does he go for a short walk once or twice a day? Is he a search-and-rescue or hunting dog, or is he a couch potato? The serious athlete may need performance formula dog food to keep up with his caloric needs—one that is higher in fat content. The less active dog probably does well on a low-fat, high-fiber diet—food that convinces him he is full even when he's not ingested many calories.

❺ Your dog's breeding should play a role in your decision about diet. Let's say you have a tiny companion such as a Yorkshire Terrier that is prone to dental disease. Well, those tiny little teeth might last a whole lot longer when you feed a diet that deters tartar accumulation. Another example: if you've taken on a giant breed, like a Great Dane, feeding adult food rather than a puppy diet during his first year of life may cut down on orthopedic issues that commonly occur during rapid growth.

If you choose to go with a homemade or raw diet, I strongly encourage you to work with a board certified veterinary nutritionist to help configure a diet that is balanced and safe for your dog (www.acvn.org). This process can often be handled over the phone or online.

❻ How much food is the right amount? Just as two people of the same age, height, and sex have vastly different caloric needs, there is no exact calorie calculation for dogs based upon breed and size. Ask your veterinarian for her recommendation, and recognize that the amount may need to be adjusted when your pup becomes thin or heavy. Note: in general, unless you have an extremely active dog, following a dog food manufacturer's recommendations results in a chubby best friend!

Where Is It Available?

Stem cell therapy is available through veterinarians who have received specialized training from the company who harvests the stem cells.

The Wave of the Future

So what does future veterinary science hold in store? Here is my best guess at what new gadgetry and wizardry will benefit dog generations to come:

PET Scans

Like the CT and MRI scan (see p. 83), the PET scan (positron emission tomography) is another way of imaging the body. During the PET scan, the patient receives an injection of a radioactive substance, and the PET scan machine detects radioactive emissions within the dog's internal tissue. In addition to providing anatomical information, such as size, shape, and location of various organs and tissues, the radioactive emissions interpreted by the PET scan also provide biological information about the cellular nature of the tissue. For example, the PET scan may be able to differentiate between benign and malignant tumors, or reveal the underlying nature of a dog's brain pathology.

More Cancer Vaccines

The year 2007 brought with it veterinary medicine's first canine cancer vaccine (for oral melanomas), and it has been a smashing success (see p. 93). Research is in the works to develop more cancer vaccines for dogs. These may completely revolutionize the way we treat canine, and we hope, human cancer.

Joint Replacements

Hip replacement surgery has worked very well in dogs. It's only a matter of time before other "bionic" replacement parts become available.

Stem Cell Therapy 2.0

Early attempts to use stem cell therapy to heal a dog's joints, tendons, and ligaments show tremendous promise. As said earlier, for dogs, stem cells hold therapeutic potential without the contentious political and religious debate that confronts their application to human medicine. I suspect other applications of stem cell therapy will take off.

"Our paradigm has been challenged, and it is gradually shifting as we look at alternative ways to select and use vaccines."

—RICHARD FORD, DVM

6 The Vaccination Conundrum

During my last year of veterinary school, I recall how scared we were when a new canine virus—parvovirus—popped up all over the country. Highly contagious, it spread like wildfire throughout the US, causing severe illness and often death. It was a downright frightening time for dog owners and vets alike. Fortunately, an effective vaccine was rapidly developed, and this horrible new virus was downgraded from a rampant deadly infection to a preventable disease.

Thank goodness for vaccines! They provide a remarkable means of both prolonging and enhancing quality of life, and have been an important component of preventive care for dogs (and, indeed, for people, too).

It's no longer in your dog's best interest to vaccinate for everything simply because you've received a reminder postcard in the mail.

Vaccinating in the 21st Century

Back in the "old" days, regularly vaccinating dogs was a no-brainer. There were only five or so vaccines to choose from, and we vaccinated every dog we could, once a year, no questions asked. My, oh my, how things have changed! Nowadays, the world of canine vaccines is a constantly shifting landscape. Figuring out when to vaccinate, and which diseases to vaccinate against, is no simple task. It's no longer in your dog's best interest to vaccinate for everything simply because you've received a reminder postcard in the mail. Just as with antibiotics, overuse of vaccines can do more harm than good. Needless vaccinations result in unnecessary expense, and every time your dog is vaccinated there is risk of serious harmful side effects. (Unfortunately, it is almost always impossible to predict which dogs will be so affected.)

Many vaccines, once routinely given annually, actually provide protection from disease for at least three years—and some for life.

Determining the best vaccine protocol requires a case-by-case assessment. Your pup will benefit when you and your veterinarian participate in this process. Consult your vet—part of your role as medical advocate—to determine which vaccinations are appropriate and when to administer them (see p. 109). Before your dog is ever vaccinated, you must feel confident that the benefits of any vaccination outweigh its risks. Herein lies the vaccination conundrum.

Consider the following facts:

- At the time of publication, vaccinations are available for 13 canine diseases.

- The duration and degree of immunity (protection) offered by any vaccine vary not only by manufacturer, but from dog to dog as well.

- It's now recognized that many vaccines, once routinely given annually, provide protection from disease much longer than one year.

In fact, the duration of immunity for most adult canine vaccines is at least three years. Some even provide lifelong protection.

- Vaccination protocols are anything but standardized. Veterinarians can give whichever vaccines they deem to be important as frequently as they see fit. Some vets give multiple inoculations at once, others administer just one at a time. Rabies vaccination is the exception in that veterinarians are required to comply with state regulations when it comes to frequency and whether it is inoculated under the skin or into the muscle.

- Increasingly clear-cut documentation shows that vaccines have the potential to cause many side effects—some life-threatening.

- The need for vaccine protection depends on where the dog lives and his lifestyle. For example, a Poodle that rarely leaves his Manhattan penthouse has no exposure to Lyme disease (spread by ticks); however, a Lab that goes camping and duck hunting may have significant exposure.

AAHA defines immunization as "a medical procedure with definite benefits and risks, and one only undertaken with individualization of vaccine choices and with input from the client."

AAHA Vaccination Recommendations

In 2003, as a result of increasing vaccination complexity, the American Animal Hospital Association (AAHA) put together a task force to come up with a set of standard recommendations for veterinarians. Task force members reconvened in 2006 to update their work. Their efforts provided vets with an excellent set of vaccination guidelines. AAHA defines immunization as "a medical procedure with definite benefits and risks, and one that should be undertaken only with individualization of vaccine choices and after input from the client." Keep in mind, AAHA has provided *recommendations* only. Veterinarians remain at liberty to do whatever they choose, i.e. which specific vaccines to give and how often.

As per AAHA guidelines, available vaccines are categorized as *core* (vaccines recommended for all dogs); *noncore* (optional vaccines that should be administered based on risk of exposure); and *not recommended* (a decision based on lack of demonstrated effectiveness or an unacceptable risk of side effects). Note: No formal recommendations are made for either rattlesnake or periodontal disease vaccinations because of "lack of experience and paucity of validation of efficacy."

AAHA outlines its recommended time interval between various vaccinations based on studies that document an increased duration of protection for some vaccines. In particular, it increases the interval for distemper and parvovirus vaccines, both of which had been administered annually. Now these two should be given no more than once every three years (see sidebar, p. 112, for details).

Despite the fact that the old standard of vaccinating annually is now deemed to be "too much of a good thing," at the time of publication, some vaccine manufacturers continue to recommend it. Therefore, veterinarians concerned they could be subject to legal action if they don't follow manufacturer label instructions may feel it necessary to continue an annual program.

Available Vaccines

Vaccines have been developed to prevent the health issues listed on the following pages. These diseases run the gamut from relatively minor health nuisances to life-threatening illnesses. As discussed,

QUICK REFERENCE
AAHA Canine Vaccine Guidelines (2006)

Core Vaccines *(recommended for every dog)*
- Canine parvovirus
- Canine distemper
- Canine adenovirus-2
- Rabies

Noncore Vaccines *(optional vaccines that should be administered based on risk of exposure)*
- Measles
- Parainfluenza virus
- *Bordatella bronchiseptica*
- *Borrelia burgorferi* (Lyme disease vaccine)
- Leptospirosis

Vaccines with Uncertain Effectiveness
- *Crotalus atrox* toxoid (rattlesnake vaccine)
- Porphyromonas (periodontal disease vaccine)

Vaccines Not Recommended *(at the time of publication)*
- Canine coronavirus
- *Giardia lamblia*

those labeled as *core vaccines* are recommended for every dog unless a negative vaccine reaction is anticipated. As you learn about the other vaccines listed, consider the possibility of exposure for your dog. This varies, depending on where you live and the kinds of activities you and your dog engage in.

Core Vaccines

- **Parvovirus** is an infection that causes severe vomiting and diarrhea, especially in puppies. Most infected dogs require several days of intensive care hospitalization for any chance of survival. Sometimes,

QUICK REFERENCE

AAHA Recommended Vaccination Schedules (2006)

Canine Parvovirus

- Puppy vaccination: begin at six to eight weeks of age, then every three to four weeks until 12 to 14 weeks old
- Initial adult vaccination (no known prior vaccinations): two doses, three to four weeks apart
- Revaccination: booster at one year of age, revaccination once every three years thereafter

Canine Distemper

- Puppy vaccination: begin at six to eight weeks of age, then every three to four weeks until 12 to 14 weeks old
- Initial adult vaccination (no known prior vaccinations): two doses, three to four weeks apart
- Revaccination: booster at one year of age, revaccination once every three years thereafter

Canine Adenovirus-2

- Puppy vaccination: begin at six to eight weeks of age, then every three to four weeks until 12 to 14 weeks old
- Initial adult vaccination (no known prior vaccinations): two doses, three to four weeks apart
- Revaccination: booster at one year of age, revaccination once every three years thereafter

Rabies

- Puppy vaccination: one dose no earlier than three months of age
- Initial adult vaccination (no known prior vaccinations): single dose
- Revaccination: booster one year following initial dose, revaccination every one to three years, depending on type of vaccine used and government regulations

Measles

- Puppy vaccination: one dose between four and 12 weeks of age to provide extra protection against canine distemper
- Adult vaccination and revaccination: not recommended

Parainfluenza Virus

- Puppy vaccination: begin at six to eight weeks of age, then every three to four weeks until 12 to 14 weeks old
- Initial adult vaccination (no known prior vaccinations): one dose is adequate

AAHA Recommended Vaccination Schedules (2006) cont.

- Revaccination: booster at one year of age, revaccination once every three years thereafter (unless manufacturer recommendations state otherwise)

Bordatella

- Puppy vaccination: one dose at six to eight weeks and one dose at 10 to 12 weeks of age
- Initial adult vaccination (no known prior vaccinations): two doses, two to four weeks apart
- Revaccination: once yearly or more often in high-risk dogs

Lyme Disease (*Borreliosis*)

- Puppy vaccination: initial dose at nine or 12 weeks of age (depending on manufacturer recommendations), second dose two to four weeks later
- Initial adult vaccination (no known prior vaccinations): two doses, two to four weeks apart
- Revaccination: annually (administer just prior to the start of tick season)

Leptospirosis

- Puppy vaccination: one dose at 12 weeks and one dose at 14 to 16 weeks of age (not recommended for dogs under 12 weeks of age)
- Initial adult vaccination (no known prior vaccinations): two doses two to four weeks apart
- Revaccination: annually

Rattlesnake Venom

- Puppy vaccination: two doses, one month apart in puppies as young as four months of age (refer to manufacturer's label)
- Initial adult vaccination (no known prior vaccinations): two doses, one month apart (refer to manufacturer's label)
- Revaccination: annual boosters timed at the beginning of rattlesnake season (refer to manufacturer's label)

Periodontal Disease (*Porphyromonas*)

- Refer to manufacturer's label

even with the best of care, the patient succumbs. It is highly contagious and transmitted via the infected dog's feces. Unfortunately, the virus is a hearty one and lasts a long time in the environment, and you can easily track it into your home on the bottom of a shoe.

Vaccination is essential for prevention of this highly prevalent, deadly disease.

- **Distemper** is a viral disease that can cause respiratory, gastrointestinal, and neurological symptoms. Often fatal, it is readily transmitted between dogs. Distemper is most commonly seen in youngsters. Heartbreakingly, even with aggressive therapy, most severely affected dogs cannot be saved.

Appropriate vaccination is highly effective at preventing distemper.

- **Adenovirus** manifests in two different varieties: Adenovirus-1 and its close cousin Adenovirus-2. Adenovirus-1 causes a fatal liver disease with secondary gastrointestinal and neurological symptoms. Adenovirus-2 is not the least bit harmful to the liver, but causes a mild upper respiratory infection. Both versions are highly contagious. Fortunately, vaccinating for either one prevents the other. The Adenovirus-2 vaccine is used because it has far fewer potential side effects than the Adenovirus-1 vaccine.

- **Rabies** inspires fear in just about everyone, and rightfully so. It infects all mammals (including humans) and is contagious between species through contact with saliva or feces (bat guano, for example). Classic symptoms, which I hope you never have to observe, are neurological and include erratic behavior, incoordination, paralysis, and, ultimately, death.

Fortunately, exposure to rabies is extremely rare. The law requires dogs be vaccinated for rabies, and local government dictates how frequently. Unlike other core vaccines, rabies must always be given separately (never as a "combination vaccine," see p. 118), and it must be administered under a veterinarian's supervision, who will provide the necessary paper record for proof of compliance.

Noncore Vaccines

- **Measles** vaccine provides cross-protection against canine distemper, and is sometimes administered to puppies less than six weeks of age because it is thought to provide better protection than the actual distemper vaccination in such youngsters. Note: It is rare that pups under six weeks of age need to be vaccinated. In kennel situations where there are large numbers of dogs, when the mother has not been vaccinated, or when the puppy leaves home at a very young age, the measles vaccine might be warranted.

- **Parainfluenza** is a highly contagious virus that is part of the kennel cough gang. Like bordatella (see below) it causes coughing and other upper respiratory symptoms. Just as with the human cold virus, parainfluenza simply runs its course and cannot be impacted by antibiotic therapy. It's a good idea to vaccinate for this disease if your dog will be in close contact with lots of other dogs. Protection from parainfluenza is often found in combination vaccines with adenovirus, parvovirus, and distemper. In addition, some kennel cough vaccines combine parainfluenza and boradatella protection (see sidebar, p. 118).

- **Bordatella** is a bacterial infection that causes a hacking, incessant, keep-you-awake-all-night kind of cough. It is cured with oral antibiotics.

Bordatella is transmitted between dogs that are housed in close quarters (hence the common term kennel cough). Although bordatella and kennel cough (specifically inflammation of the windpipe and lower airways) are often referred to interchangeably, keep in mind that bordatella is only one of the infectious organisms capable of causing kennel cough, thus limiting the bordatella vaccine's protective abilities. Bordatella is commonly combined with parainfluenza in kennel cough vaccines (see sidebar, p. 118).

Many boarding kennels require guests to be vaccinated for Bordatella. Not only do they want your pup to return home without a cough, they want to feel confident that he isn't a "Typhoid Mary," infecting every other dog staying there. Note: If you are vaccinating your dog for boarding purposes, be sure to do so within the appropriate timeframe. It takes a minimum of two to five days for the vaccine to produce immunity (depending on whether the vaccine is administered via injection or directly into the nasal passageway). If your dog is vaccinated the day before his stay away from home, the kennel proprietors may be satisfied, but you won't have significantly reduced your chances of bringing home a coughing dog!

- **Lyme disease** has received a lot of press because it can have such devastating effects on people. The evildoer in this disease is a tick-transmitted *spirochete*, a bacterium that is shaped like a corkscrew. The most typical symptom seen in dogs is sore joints manifested as stiffness, reluctance to get up and move about, and lethargy. Less commonly, inflammation in other tissues such as the brain, heart, or kidneys occurs. When caught early, Lyme disease can often be cured with a course of antibiotics. Just as in people, the more chronic the disease, the more difficult it is to cure. The vaccine isn't foolproof but should be considered if your pup has significant exposure to ticks.

- **Leptospirosis** is a bacterial infection that causes kidney failure. Typical symptoms include lethargy, loss of appetite, and vomiting. Liver damage may occur as well. This organism thrives in the urine of wildlife and livestock without causing harm to them. Unfortunately, this is not the case in dogs—their kidneys won't stand for it.

The good news is that leptospirosis can often be cured with antibiotic therapy. The key is early detection and diagnosis. Exposure clearly depends on the dog's lifestyle and home environment. If your pup never leaves your well-manicured condominium complex, no need to worry. However, if you reside on a cattle ranch or your dog is lucky enough to accompany you on hiking and camping trips, consider the vaccine. Rely on your vet to let you know if leptospirosis is prevalent where you live and travel with your dog.

This is often found in combination vaccines that also contain protection against adenovirus, parvovirus, and distemper.

Vaccines: No Recommendation (Uncertain Effectiveness)

- **Rattlesnake venom** is nasty business. Unfortunately, some dogs (famously those of the Jack Russell Terrier variety) relish the opportunity to play with a serpent! The good news is, rattlesnake bites rarely prove fatal (in dogs or humans) when appropriate therapy (antivenin, antibiotics, intravenous fluids) is provided.

At this time, the jury is still out on whether or not the rattlesnake vaccine is truly effective. Bitten dogs exhibit painful, rapidly progressive facial swelling at the site of the bite—while the vaccine doesn't prevent symptoms from occurring, it is possible it lessens their severity.

The available vaccine is intended to protect dogs against the venom associated with the bite of the Western Diamondback, and

QUICK REFERENCE

Combination Vaccines

Although rabies is *always* given as a separate injection, other vaccines are often combined (your dog receives only one rather than multiple shots) for convenience and comfort. The DAPP vaccine is the most common combination vaccine. It combines protection against four different diseases: distemper, adenovirus, parvovirus, and parainfluenza. The kennel cough vaccination often combines protection against parainfluenza and bordatella.

possibly the Eastern Diamondback. At the time of publication there is no evidence of protection against venom from the Mojave Rattlesnake.

- **Periodontal disease** refers to gum inflammation, dental tartar, infected teeth, and bad breath. It is not contagious. If you share your life with a miniature breed (notoriously predisposed to periodontal disease), you are likely quite familiar with this syndrome.

The periodontal vaccine targets three types of bacteria that may predispose a dog to periodontal disease. At this time, it is uncertain whether or not this vaccine truly provides significant protective benefit.

Vaccines: Not Recommended

- **Giardia** is a highly infectious disease caused by a protozoal organism that can cause diarrhea, vomiting, loss of appetite, and weight loss. (It is what people worry about getting when drinking unfiltered river, stream, or lake water.) Giardia is readily transmitted between dogs and many other species, including humans. A course of oral medication typically resolves the infection. At the time of publication, the giardia vaccine has not been shown to prevent infection.

- **Coronavirus** can cause vomiting and diarrhea. It is spread between dogs via infected feces. The vast majority of infected dogs show absolutely no symptoms. This vaccine is "not recommended" because of low disease prevalence and the mild self-limiting nature of symptoms when they do occur.

Vaccination Adverse Reactions

All 13 canine vaccines on the market today have received FDA (Federal Drug Administration) approval. Nonetheless, just as in human medi-

MISSY'S story is a tough but important one for me to tell. Picture a Bichon Frise—a pristinely white, cottony fluff ball of a dog. Missy was adorable and adored, with gorgeous brown eyes, an incredibly sweet disposition, and a little red bow atop each ear. She was eight years old when I first saw her, and she had never before experienced a medical problem of any kind.

When I met her it was two weeks after she had been vaccinated for Lyme disease and leptospirosis, and she was suffering from a life-threatening side effect. One or both of these vaccines had managed to trigger Missy's immune system so that it attacked and destroyed her own platelets—blood cells necessary for normal blood clotting to occur (known as *immune mediated thrombocytopenia*). She was bleeding internally. Though we tried to stop the bleeding with transfusions, and gave her medication to tame her overly active immune system, we lost the battle, and Missy succumbed to the disease.

Everyone was deeply saddened by her untimely death. In addition to sadness, I felt outrage that it had been so utterly and completely unnecessary. You see, had anyone paused to think for a moment, she should never have received the vaccines she did. In fact, Missy's pristine and sheltered indoor lifestyle completely prevented exposure to either the ticks that transmit Lyme disease or wildlife that pass on leptospirosis. Missy and her family paid the ultimate and tragic price for such nonsensical decision-making.

cine, every once in a while adverse side effects occur. Sometimes the reaction occurs immediately following the injection, sometimes not until days, or even weeks, later. Most are the result of the dog's immune system "overreacting." Symptoms can be as mild as some transient swelling of the face, or as serious as *anaphylaxis*—a life-threatening allergic reaction resulting in collapse and labored breathing—and the reaction that Missy experienced (see sidebars, pp. 119 and 121). When serious, it can be heartbreaking for everyone involved, and even more so when the vaccine wasn't really necessary in the first place.

What to Do When Adverse Reactions Occur

Veterinarians are encouraged, but not required, to report adverse vaccine events to the vaccine manufacturer and the Center for Veterinary Biologics, located in Ames, Iowa. This is for purposes of collecting data so vets and manufacturers can continue to learn more about vaccinating dogs. So, for the sake of all "dogdom," if your pup has a bad reaction, let your vet know you would like it officially reported.

Revaccination Decisions

When your dog has experienced a severe or life-threatening vaccine reaction, you should strongly consider foregoing vaccinations in the future. Even if a reaction was mild, careful consideration is a must. Vaccine serology, also known as "vaccine titers" or "titer testing" (see p. 122) helps you determine if your dog is truly in need of another vaccination. If you choose to revaccinate, your pup can be given a different brand of vaccine, and medication (such as an antihistamine) in advance to possibly prevent another adverse reaction. It is sometimes advisable to avoid giving your dog the combination vaccinations that come in one shot (see p. 118). Instead, give the vaccines separately, spread out over time.

QUICK REFERENCE
Potential Adverse Reactions to Vaccinations

Injection Site Reaction
Abscess formation
Hair loss
Itchiness
Lump or tumor
Pain
Swelling

Intranasal Vaccination Site Reaction
Coughing
Discharge from the eyes
Sneezing for three days or longer
Ulcers in the nose or mouth

General Reactions
Anaphylaxis (life-threatening allergic reaction)
Autoimmune diseases involving blood cells, kidneys, or joints
Behavior changes
Decreased energy
Difficulty breathing
Fever
Jaundice
Kidney failure
Neurological abnormalities
Swelling around the face or ears
Vomiting

Note: If you and your vet decide to discontinue all vaccinations, ask your vet to write a note of explanation to your local government agency excusing your dog from future rabies shots. Unfortunately, even with a note, many boarding facilities won't budge on the requirement that their guests be vaccinated for "kennel cough."

Vaccine Serology

Okay, dear reader, prepare yourself for what may be the driest, most tedious section of this book. A bit of immunological science is necessary because I want you to understand the concept of *vaccine serology*—a way of testing your dog's blood to determine whether or not he is in need of revaccination. The duration of vaccine-induced protection is highly variable from dog to dog: a distemper vaccine that protects my dog for eight years might only last four years in yours. How can you deal with this uncertainty? There are two choices. You can vaccinate at recommended intervals whether you're sure your dog needs it or not, or you can ask your vet for serology.

The component of the blood examined is called *serum*—hence the term "serology." When vaccine serology for a specific disease is performed, antibodies—microscopic immune system "soldiers"—are quantified. The results are referred to as "antibody titers." A *negative* or *low titer* means that your dog is *not* adequately protected and revaccination is indicated. A positive or "protective" titer indicates an adequate number of antibodies and means the dog is protected and revaccination is unnecessary.

Serologic or "titer testing" provides a stellar means of avoiding over- or under-vaccinating dogs.

Vaccine serologic testing is rapidly growing in popularity. Also known as "titer testing," it provides a stellar means of avoiding over- or under-vaccinating dogs. At the time of publication, testing is available for parvovirus, distemper, and rabies. All that is required is a small blood sample from the dog.

Vaccine serology is most commonly used in the following three situations:

Puppies

It is currently recommended that puppies receive their first vaccinations at six to eight weeks of age, followed by revaccination every three to four weeks until 12 to 16 weeks of age (see p. 123 for a complete list). This usually translates into three sets of shots, but they need *at least* two to induce adequate protection.

Why so many puppy vaccinations? A short lesson in vaccine physiology will help you understand the reasons why.

Through nursing, a puppy receives disease protection for eight to

18 weeks after birth. Referred to as "maternal immunity," this protects the pup from getting the same diseases we hope to later prevent with vaccinations. An "army" of mom's antibodies is deployed to attack and neutralize the infectious organism after it enters the pup's body. The problem is that any vaccinations we give will be neutralized (rendered a "placebo") as well if maternal immunity is still "on guard." Let's say that a pup is vaccinated for distemper and parvovirus at eight weeks of age, but his maternal immunity doesn't disappear until nine weeks. The vaccine administered at eight weeks will have been useless—hence the need to repeat vaccinations every three to four weeks until the pup is 12 to 16 weeks old to be sure he is protected.

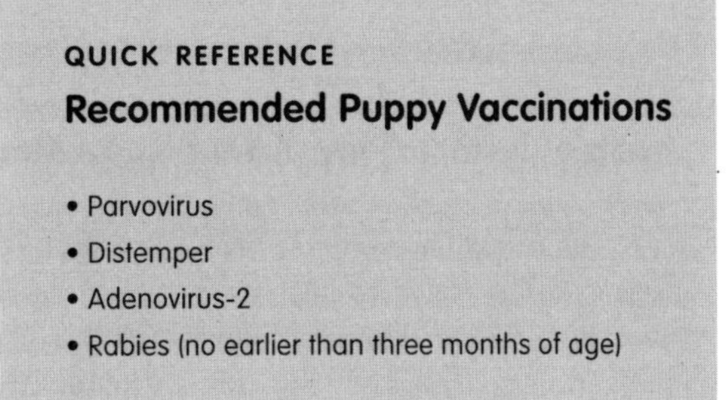
QUICK REFERENCE

Recommended Puppy Vaccinations

- Parvovirus
- Distemper
- Adenovirus-2
- Rabies (no earlier than three months of age)

So, how does serologic testing fit into this picture? It fits very well! Many veterinarians will perform serology two weeks after the last official round of puppy vaccines (typically at four months). If the titer is negative or too low to be protective, this means that maternal immunity was long-lasting and all of the vaccines were neutralized. Another possible explanation is that the particular type of vaccine used didn't trigger the pup's immune system, and a different brand of vaccine might be considered. Either way, another round of puppy vaccines is needed. The vet will continue to vaccinate and hope that the maturation of the dog's immune system, or the alternative brand of vaccine, gets positive results.

Adult Dogs

In adult dogs, the duration of immunity from a vaccine is highly variable. In some cases, puppyhood vaccinations last for life, while other

SECRET FOR SUCCESS

Keep a Current Copy of Your Dog's Medical Record

I encourage you to keep a current copy of your dog's complete medical record (doctor's notes, official reports, and all laboratory data) with all of your other important personal documents. No, I am not forecasting that your local veterinary clinic will burn down, or that your vet's hard drive will explode. There are many reasons why the savvy medical advocate would keep one around (by law, you are fully entitled to a copy of your dog's medical record).

- You never know when the need for an emergency visit will arise. If it's in the middle of the night or on a Sunday, you may have to take your dog to an unfamiliar emergency facility, or the vet on call may work at a clinic other than your own. Neither will likely have access to your dog's medical record—a huge handicap, especially when time is of the essence. Perhaps the vet needs to use anesthesia to sew up a laceration, but you recall that when he was anesthetized a couple of years ago, the vet told you his heart stopped or his breathing ceased, or something terrible like that happened. It is imperative the current doctor be privy to these details.
- Most veterinary clinics and hospitals purge medical records that have been inactive for three years, and what happened years ago can sometimes significantly impact a veterinarian's way of thinking about her patient's current issues.
- If you travel with your dog (lucky dog!), taking along a current copy of his medical record can again save time and trouble should an unplanned trip to the vet be necessary.

Please don't feel awkward asking for copied updates of your dog's medical record; veterinary staff are used to this request. Simply explain why you want a copy. Provide reassurance that you are not "jumping ship" (unless, of course you are, in which case an honest explanation will suffice). Don't be surprised if you are charged a fee for medical record duplication.

Also, just in case something terrible does happen (here in California, you never know when the "big one" is going to hit), remember it's always prudent to have more than one copy of an important singular document, whether yours or your dog's.

dogs will need ongoing vaccinations. In order to avoid over-vaccinating and potential side effects (see p. 121), you can opt for serologic testing. When your dog is found to have adequate antibodies, you can forego revaccination for the year. If the serology results demonstrate that the antibody titer is negative or too low, revaccination is justified.

Dogs with Previous Vaccine Reactions

Whenever a dog has had an adverse reaction to a vaccine, there is always the potential for a repeat performance. One is left with the question of which would be worse: another vaccine reaction (often more serious the second time around), or the disease itself. Serologic testing helps sort through this dilemma. If it shows that the dog has adequate protection—Whew! Another vaccine and its potential side effects can be avoided. If testing indicates disease protection is lacking, it's time to have a lengthy conversation with your vet about the pros and cons of revaccination.

Talk to Your Vet before Vaccinating

As you now likely realize, there's a good deal to discuss with your veterinarian before your dog is vaccinated. Here is a summary of the steps to take when assessing the potential risks and benefits of suggested vaccinations:

1. Begin by figuring out which diseases your dog has potential exposure to. For example, if your dog is frequently boarded, the Bordatella vaccine makes perfect sense; if you take your pup hiking and camping, then Lyme disease should be considered.

2. Alert your veterinarian to any recent symptoms or medical issues your dog has experienced. Vets prefer to avoid vaccinating a dog

QUICK REFERENCE

Vaccine Issues to Discuss With Your Veterinarian

❶ Potential for disease exposure
❷ Current medical issues
❸ Prior adverse vaccination reactions
❹ Serological testing ("titer testing")
❺ Potential vaccination side effects

that is sick, as it is better to let his immune system concentrate on getting rid of the current illness rather than creating a vaccine "distraction." If your dog has a history of autoimmune disease—such as glomerular disease (p. 368), autoimmune hemolytic anemia (p. 312), immune mediated thrombocytopenia (p. 314), immune mediated polyarthritis (p. 343)—it may be advisable to forego ongoing vaccinations because of the potential for triggering disease recurrence.

❸ Next, be sure to let your vet know if your dog has had vaccine side effects in the past (see p. 121). If the reaction was quite serious, she may recommend that you forego future vaccinations, necessitating an official letter to your local government agency excusing your pup from rabies-related requirements.

❹ Ask whether serologic testing makes more sense for your dog than simply vaccinating at set intervals.

❺ Finally, inquire about the potential side effects of proposed vaccinations, what you should be watching for, and whether or not

QUICK REFERENCE

Vaccinating During Pregnancy

In the past, vaccinating a pregnant dog was considered taboo due to lack of clear-cut data concerning safety and effectiveness. According to the Centers for Disease Control and Prevention, vaccination risks to a pregnant woman and her developing fetus are only theoretical, and it is advised vaccines be administered when needed. Further research is needed in veterinary medicine, but the current wisdom is to extrapolate what is known about pregnant women. The risks associated with leaving mother (and therefore unborn pups) unprotected are felt to outweigh the risks of vaccinating her during pregnancy. My hope is that your dog is up-to-date on her vaccines before she conceives. But, if you find yourself with an unvaccinated, pregnant dog on your hands, talk to your veterinarian about the types of vaccines best suited for her.

there are any restrictions for your dog in the days immediately following vaccination.

Vaccination Clinics

I will tell you right up front that I am not a fan of vaccine clinics—a "factory line" approach to vaccinating dogs. They can be sponsored by veterinary hospitals or mobile vaccination services that set up shop in pet stores, feed stores, and other animal related locales. Their only redeeming quality seems to be the low cost that makes it possible for some dogs to be vaccinated that otherwise wouldn't be. Know that, if you choose to use a vaccination clinic, you may be sacrificing quality of care for your dog in the following ways:

1. You may not receive adequate counseling about which vaccinations are appropriate for your dog, based on his age and lifestyle.

If any counseling occurs, the conversation will likely be brief and may not be with a veterinarian.

❷ Serologic testing is likely not an option.

❸ A thorough physical examination will not be performed prior to vaccination administration. Abnormalities such as fever, irregular heart rhythm, or an abdominal mass will go unnoticed. Not only might the vaccination do more harm than good in a dog that is sick, but a golden window of opportunity for early disease detection and treatment will be missed.

❹ Records pertaining to prior vaccination reactions may not be available. In addition, the vaccine clinic might not document all of the information that is an important part of your dog's current medical record. This should include the date of vaccination, vaccine manufacturer and serial number, method of injection (under the skin, in the muscle, in the nostril), and identity of the vaccinator. Such information is especially important in the case of an adverse vaccine reaction.

❺ The vaccination clinic vet may not be available to tend to your dog if he experiences an adverse reaction, especially one that occurs hours to days later.

Vaccinating Your Own Dog

Sometimes, for efficiency or to save money, people want to administer vaccinations themselves. Often these are people who live with a "herd" of dogs rather than just one or two. If you fall into this camp, do your best to not compromise your dogs' best interest by meeting the following criteria before vaccinating:

1. Your dog appears healthy, and your vet has given him a clean bill of health based on a recent—within the past month—physical exam.

2. You've consulted your vet about potential risks and benefits of the vaccinations that are appropriate and due.

3. You've purchased the vaccines from your vet or a reputable pet store, feed store, or online source. Note: When using mail order, be sure the vaccines arrive on ice—those at room temperature when received should be returned. Check the expiration dates on the vials before using them.

4. You've been properly trained on how to give vaccinations and are clear about which ones go where.

5. You are well versed in adverse reactions and have ready access to your veterinarian should such a reaction occur. (In other words, don't vaccinate on Sundays or in the middle of the night!)

6. You keep detailed medical records including the vaccine's name, manufacturer and serial number, as well as the date, site, and route of administration. Many vaccine vials have labels that double as stickers to place within the medical record.

"Better to ask twice than to lose your way once."

—DANISH PROVERB

7

Important Questions to Ask Your Vet...and How to Ask Them

You're listening to your veterinarian provide a rather long-winded dissertation (using words learned only in veterinary school) about the canine love of your life. You were worried to begin with; now, as your vet is describing the five-syllable diagnosis and various treatment options, your concern is escalating into full-blown anxiety. You're actually not hearing much at all because all of your brainpower is hard at work keeping the tears at bay.

When your vet comes up for air and asks if you have any questions, you don a fake half-smile and shake your head to say, "No," when what's screaming inside your head is, "Questions! I don't know enough to have a clue as to what questions I should ask!" Your vet may think that she's just provided you with a whole bunch of useful information, but all you heard is that the wiggling and wagging center of your universe is sick. It's tough to ask meaningful questions when you've just been handed news that leaves you emotionally "unglued."

The recipe for successful medical decision-making calls for a pinch of gut instinct combined with a whole lot of clarity about the potential risks and benefits of the available options.

SECRET FOR SUCCESS

Bring Along a "Hired Gun"

When your dog has a significant health problem and you have scheduled a visit to the vet, consider bringing along an extra set of eyes and ears (those of the human variety work best); someone you trust to give you objective feedback rather than someone interested in persuading you to do what they think is best.

Just like static on the radio, "emotional static" can significantly interfere with your ability to truly hear what your veterinarian is saying, or to make significant decisions when the news is complicated and worrisome. Your "hired gun" should be a good listener, ideally have some medical savvy, and should be capable of remaining emotionally grounded with respect to your dog. His or her job is to pick up on all the details you may miss, ask questions, and debrief you after the visit—not to make decisions on your behalf.

Even if you are able to quell your emotions, coming up with significant questions isn't necessarily easy. After all, how can someone with no medical training know what to ask, especially when it is unclear just what information you truly need? Yet, here I am telling you that such questions are *essential*. This is the reason why: the recipe for successful medical decision-making calls for a pinch of gut instinct combined with a whole lot of clarity about the potential risks and benefits of the available options. Getting the answers you need from your vet is clearly the best way to understand your dog's specific condition and to make the right choice for him. In your role as "canine medical advocate extraordinaire," I have no doubt that you will choose clarity over guesswork any day of the week.

I'm going to make it easy for you! This chapter is filled with questions to ask your veterinarian on the general topics most likely to affect the majority of dogs. In addition, I've provided a comprehensive list of disease-specific questions in the Appendix, p. 283. I've included the questions I believe are essential for getting the information you need to make good choices about your dog's medical care. And, I encourage you to add your own, based on your dog's circumstances or special needs.

65% of dog owners include their pups in Halloween festivities.

So, when exactly are you supposed to ask your veterinarian these important questions? When you've just received your pup's diagnosis, I'm afraid you cannot expect your vet to sit patiently by while you pull out this book! Here's my recommendation: whether you use my suggestions (from both this chapter and the Appendix, p. 283) or come up with your own, vets recognize that clients need time to process new information and put together their list of questions before making significant decisions. You may feel completely comfortable making some decisions during your first office visit, so ask your veterinarian to get started on the choices about which you are certain. Vets anticipate the need for more discussion later, either in person, over the telephone, or via email.

General Recommendations

1. Prepare a *written* list of your questions. You may normally have a crackerjack memory, but it's very easy to become forgetful or distracted under the influence of considerable emotional energy and anxiety.

2. Do your best to listen to your veterinarian's explanation *first*. If she is a thorough and effective communicator, she may answer most, if not all of your questions before you've even had the chance to

ask them! Your written list will allow you to make sure that nothing has been missed.

❸ Consider writing down, or better yet, audio-recording your vet's responses. This allows you to review the information and can make it easier to solicit opinions from other people you trust. Even the most patient veterinarian gets a bit "tweaked" when she has to answer the same question over and over again—whether for you or for others who weren't there for the initial conversation.

❹ Don't hesitate to ask for unfamiliar words to be pronounced again, or better yet, spelled out. We vets often get so busy describing things, we sometimes forget which words are "medical-speak" and which are "lay-speak." As described in chapter 4, always "pause for confusion"!

❺ Don't allow yourself to be distracted when listening to your vet's responses. I know it's not always easy, but try to suppress any strong emotions capable of derailing your train of thought. Sometimes, it's better to have your dog out of sight when you are asking your questions. His antics have the potential to interfere with your concentration.

❻ If you are worried about the financial aspects of your choices, refer to chapter 10 for assistance in formulating money-related questions.

❼ If you get the vibe that your veterinarian is "running on empty" in the patience department, I beseech you, please don't feel intimidated and rush into a decision before *all* of your concerns have been addressed. She may simply be behind schedule with two emergencies sitting in the waiting room. If so, perhaps it would be better to arrange a telephone or email conversation at the end of her workday, or schedule another appointment for further discussion. Just ask!

Universal Questions

A few questions are relevant no matter what problem your dog is experiencing. They also happen to be the ones easily overlooked. Here are my "Top Five" in this category:

❶ What should I be watching for?

Whenever your dog has been diagnosed with a disease, ask what symptoms you should watch for; when he is started on a new medication, ask about the possible side effects. What should prompt you to call your vet, have your dog reevaluated, or pay a visit to the emergency hospital?

I'll never forget a particular client's recounting of the inadequacy she experienced after finding out her dog had serious heart disease. My client assumed that if her dog was in trouble, he would appear to be having a heart attack (i.e., he would simply "keel over" like one of her relatives had). No one ever explained to her that dogs don't do this. Rather, their breathing becomes labored because of fluid accumulation in or around the lungs. She was so preoccupied watching for signs of a "heart attack," she discounted her dog's struggle to breathe until it was too late to help him.

SECRET FOR SUCCESS

Be Prepared for an Emergency

Thanks goodness for emergency care facilities—dogs just love getting into trouble on Sundays and holidays when the family vet's office is closed! Some small animal vets take their own emergency calls or share being "on call" with a few others in their area. But more and more vets have come to rely on regionalized emergency facilities that are open evenings, weekends, and holidays. Some are also available during weekday hours to provide care if your clinic happens to be closed for lunch or their schedule is too busy to accommodate an emergency. Not only can veterinarians send their after-hours emergencies there, they can also refer patients requiring 24-hour care. Ask your family vet how emergencies should be handled.

The fact that your dog is in need of emergency care is in and of itself stressful, but going to a strange hospital can be cause for further anxiety. There you will be working with people you don't know, other emergencies may need attention before your own, fees may be higher than what you are used to, and if your dog is in dire straits, he may be whisked away from you before you've even had a chance to talk with the doctor! However, if you're prepared before the event, you will find it much easier to regain some control over the situation even when it feels like it is out of your hands.

❶ Know where you are going! Right after Manny ingested a toxin or Peanut fell off the back deck is not the time to be looking through the phone book for the nearest emergency care facility. Always have the telephone number and address of a recommended hospital readily available (and be sure your pet sitter has this information as well). Do a couple of practice drives to the clinic location so you know the quickest route.

❷ When you are unsure of the urgency of a situation, call the clinic staff. If, based on your description, they are also uncertain, pack your pup into the car— it's better to be safe than sorry.

❸ What happens when you come home from work to find a chewed-up, empty container of grandma's "water pills," and two dogs with suspiciously innocent expressions on their faces? Which dog enjoyed the container and which enjoyed the medication is anyone's guess! The bottom line? If you are unsure about who ingested the substance, whether or not the substance is toxic, and how long ago it happened, call your emergency veterinary contact or the ASPCA Animal Poison Control Center (888-426-4435) to find out how best to proceed. If your dog has swallowed something poisonous or toxic, it's an emergency. The sooner treatment is initiated, the better.

❹ If your dog did ingest something potentially toxic, bring it along, and when applicable, the container it was stored in.

❺ If you can't find someone else to take you and your dog, drive safely. Yes, your dog may be having a seizure in the back seat of the car, and you might not even be sure whether or not he is still breathing. As difficult as it may be, if you are alone, you must give your full attention to the road.

❻ If possible, bring along your dog's medical record. As described earlier, it might prove invaluable to a new vet.

❼ Don't assume you must make rushed decisions. Ask the veterinarian what you need to decide right away. Then, spend time weighing your options in terms of nonurgent care.

Here's an example: I recently received an emotional telephone call from my cousin. She was at an emergency clinic because a car had rolled over her elderly Golden Retriever's leg. Fortunately, Pia was just fine other than some broken metatarsal bones (small bones in the lower hind leg), which the veterinarian strongly recommended be surgically repaired. The alternative was a simple splint, which the vet thought would likely lead to chronic pain within the affected leg. My poor cousin was beside herself trying to figure out how to proceed. She had strong reservations about surgery given Pia's advanced age.

I suggested that my cousin did not need to tell the vet her choice at that very moment. In this case, her dog would be just fine with some pain medication and her leg in a splint while my cousin took the time needed to properly research and consider her choices. She consulted with her family veterinarian as well as with a surgical specialist, and in the end, she opted for the more conservative, nonsurgical approach. Four months later, her "golden oldie" was comfortable and walking remarkably well.

❽ As the story above shows, just because, you are in an emergency situation, does not mean you cannot educate yourself about your dog's condition. Call a friend or relative who is not under the influence of adrenaline to serve as your sounding board. Ask the veterinary staff members if they can supply you with some reading material or Internet access for on-the-spot research.

❷ What is the prognosis?

This question asks your vet to give you a sense of what the medical outcome will be for your dog. She may be able to provide you with statistics such as, "We expect complete recovery 80 percent of the time," but if this kind of data is unavailable, she may prognosticate based on her clinical experience. Whatever information you receive will help you prepare from both a practical and an emotional point of view.

❸ What should I do differently at home?

Your veterinarian may recommend some lifestyle changes for your dog. For example, if he has a contagious disease or is taking medication that suppresses his immune system, you may be asked to keep him isolated from others. An increase or decrease in his normal exercise regimen is sometimes recommended. If your dog shouldn't be jumping up or down, you may be instructed to set up a ramp or set of steps to get in and out of the car, or on and off your bed.

❹ What should I do in case of an emergency?

Every veterinarian has her own preferences for handling emergencies (see p. 39). In the daytime, she may prefer you call first, or her policy may be to just bring your dog right in (be sure that the office doesn't close for lunch!) Your vet may take her own after-hours emergencies, share all or some with an on-call group of doctors, or refer her clients to a particular emergency hospital. Know where you can go and who you can call anytime—night or day.

❹ What happens next?

Any time you talk with your veterinarian, before you leave her office or hang up the phone, *know what is supposed to happen next*. Perhaps she would like you to call her in a couple of days or weeks. Maybe she wants you to

bring your dog back to her office for a follow-up visit in four months. Or, your dog—in excellent health—may not need to return until it's time for his next annual physical examination.

Here's a classic example of what can happen when the next step *isn't* discussed. Say your dog is predisposed to the formation of bladder stones—little rock-like structures within the urinary bladder that cause blood in the urine, straining to urinate, and bladder infections. Surgery is necessary to remove the stones from the bladder—a pretty big deal for your dog (and for your finances). Once removed from the bladder, the stones are analyzed for their mineral content. This analysis helps your vet figure out how to prevent the stones from reforming, with options such as changing diet, prescribing medication, and increasing water intake. Now, the *key* to preventing more stones is to perform follow-up urine tests to be sure that the preventive therapy is working. But unfortunately, as happens all too often, these tests and the necessary information they provide are lost in the shuffle.

Wouldn't you know it—a year down the road, your dog develops more bladder stones and surgery must be repeated. What a disservice to your poor pup! To prevent this sort of thing from happening, never let the conversation end without asking, "What happens next?"

Questions on General Topics

On the following pages, I address:

Questions about a Newly Prescribed Medication

I grit my teeth when talking to some of my relatives about the medications they are taking! While they can usually describe the size, shape, and color of the pills they are swallowing, they often cannot tell me the actual names of the drugs, potential side effects, or even why they've been prescribed. When it comes to medicating your best little buddy, don't let this be you! Get answers to the questions listed below before administering a single pill or capsule.

- **What is the medication supposed to do?**

Perhaps this sounds like a ridiculous question, but I see many clients who give their dogs a handful of pills every day without any certainty of what they are for. Write down what the medicine is for, and how it works.

- **What are the potential side effects, and what should I do if I observe them?**

Your veterinarian will be able to provide you with a list of possible side effects. If any are observed, depending on the drug, she might ask you to skip a dose, decrease the dosage, or discontinue the medication altogether. She will want you to give her a call and let her know how the change worked. In addition to the "typical" side effects, keep the following disclaimer in mind: any medication in any individual is capable of causing an *idiosyncratic reaction*. Should your dog exhibit something odd or different shortly after beginning a new prescription, it is reasonable to think it may be a side effect to that medication. Report this to your vet.

- **Are there any special instructions?**

Some medicines are better absorbed or tolerated on an empty stomach, others need one that's full. Some must be shaken, refrigerated, or kept in a dark cabinet. These details are important—you don't want the drugs rendered ineffective.

- **Is the medication compatible with others my dog is taking and can they all be given at the same time?**

Many drug combinations are perfectly compatible, but others downright toxic. Sometimes, a particular drug interferes with the normal absorption of another, so they should not be given at the same time. Any new medication should be assessed for compatibility with all current prescription and over-the-counter medication, herbs, vitamins, homeopathic remedies, and supplements. Your doctor will love you if she is provided with a *written* rather than a *verbal* list of everything your dog is currently receiving.

- **Does the timing need to be exact?**

If you are asked to give the drug once a day, does it matter what time of day? Similarly, if you are instructed to give the drug three times a day, is there a little bit of leeway, or must doses be administered exactly eight hours apart? I've seen clients rearrange their entire lives (unnecessarily) to do exactly as the prescription label is written.

SECRET FOR SUCCESS

Learn as Much as You Can about Your Dog's Medications and Supplements

Just as I hope you wouldn't "pop a pill" or take a supplement without knowing all that it has the potential to do (good and bad), I encourage you to be equally vigilant about what you give your dog. Rely on your vet and pharmacist (many products for dogs are available at regular drugstores) to explain the benefits of the medication or product, as well as potential side effects and adverse interactions with other drugs. Make sure your advisors are aware of everything else your dog is receiving (some drugs, supplements, and herbs don't "play" well together).

Cross-reference what you are advised with what you read. Invaluable resources are current editions of the *Physicians' Desk Reference (PDR)* for human drugs and Donald Plumb's *Veterinary Drug Handbook* for veterinary pharmaceuticals. In addition, there is also a *PDR* for nonprescription drugs, dietary supplements, and herbs (the *Physicians' Desk Reference for Nonprescription Drugs and Dietary Supplements*). These books describe how a product works, reasons for use, storage recommendations, warnings, adverse side effects, potential interactions with other products, and dosage information.

Negative Drug Side Effects

I don't have to convince most people that drugs can have negative side effects, but I do sometimes have to twist some arms to convince clients that the same is true for herbs and supplements. The loud and clear truth of the matter is any supplement, herb, and drug is capable of causing a negative or idiosyncratic (peculiar) reaction in any given dog. Although milk thistle (an herb used to promote liver health) isn't supposed to have any side effects, if your dog begins vomiting within days of beginning its use, you must consider the milk thistle to be a possible cause.

Familiarize yourself with the potential side effects of anything your dog is taking. If you are giving your dog prednisone, and his panting wakes you up in the middle of the night, you'll be able to relax, knowing this is a common prednisone side effect (and perhaps you'll be less inclined to throw a pillow at him for keeping you awake). Trust me when I say you will be spared needless worry when you not only know which side effects might arise, but also which ones are "harmless," and which are cause for concern.

Negative Drug Interactions

A drug, supplement, or herb that is perfectly harmless by itself, can be capable of wreaking havoc when combined and ingested with other drugs, supplements, or herbs. For example, an older dog that has been doing just fine for years on daily nonsteroidal anti-inflammatory medication to treat his arthritis pain may develop severe, potentially life-threatening stomach ulcers if oral cortisone is added to the mix.

Avoid Prescription Errors

Just as is true for your own medication, it's important to be on the lookout for prescription errors. So many little details go into accurately prescribing and filling a prescription, as much as everyone tries to avoid mistakes, they occur with alarming frequency. Can the staff member filling the prescription (at the veterinary office or commercial pharmacy) read the vet's messy handwriting? Is the correct drug grabbed off the shelf? Are the prescription label's instructions typed accurately? With so much potential for error, how in the world can you be an effective "watchdog" for your dog? Here are some suggestions.

❶ Pull out your glasses so you can read all the tiny writing on the label. By law, every prescription label must contain very specific information to help you verify it was filled correctly. See p. 145 for details.

❷ Verify that the strength of the medication and instructions for administration agree with what your vet told you. If she said to give a liquid twice a day, but the prescription contains tablets to be given three times a day, reconcile this difference.

❸ When medicine is refilled, compare its appearance to what you gave before. If it is not the same size, shape, and color, ask questions.

❹ Bring along all of your dog's medication to every veterinary visit. Your veterinarian can review the prescriptions and detect errors or inconsistencies by comparing them to the medical record. With the bottle in hand, it's a whole lot easier for your vet to verify what your dog is actually receiving than it is for her to interpret, "I give him half of the small yellow one twice a day." Veterinarians manage to keep a lot of information in their heads, but they don't have room to remember the shapes, color, and sizes of all the medication they prescribe.

❺ If a pharmacist suggests a generic form of a drug to decrease your expense, verify with your veterinarian that this is an acceptable substitution. Just because the generic works well in a human, doesn't mean it will in your four-legged friend.

Where to Purchase Your Dog's Prescription Medications

When purchasing prescription medications for your dog, you have three sources: your veterinary clinic or hospital, a regular commercial drugstore, or an online pharmacy.

Veterinary Clinic or Hospital

This first option is not only convenient, it also has significant built-in quality assurance. Office staff members should be familiar with the nuances of veterinary pharmaceuticals. They inspect the drug to be sure it has been handled and shipped properly. For example, when a medication must be refriger-

SECRET FOR SUCCESS CONT.

ated, they notice if it arrives as it should, or if it is frozen or warmed to room temperature. There are also fewer opportunities for prescription error as only two people are involved in the process—the veterinarian who writes the prescription and the staff member who fills it. The only downside can be cost, as you might find the same medication is cheaper elsewhere.

Commercial Pharmacy

Unless the drug prescribed is strictly a veterinary pharmaceutical, human drugstores can fill your dog's prescriptions, just as they fill your own. In fact, most of the drugs veterinarians prescribe are used for people as well, although often for different reasons. For example, Viagra® is used in dogs to treat a disease called pulmonary hypertension (a lung disease that secondarily causes heart problems). Clearly, this drug is dispensed to humans for an entirely different reason!

A significant benefit to purchasing from a drugstore is that some offer compounding services—they will take a "human-size" drug and crush, pulverize, mix, and otherwise transform it into a doggie-sized dosage. Some are even willing to formulate the drug into a tasty meat-flavored chew treat or liquid. In some cases a trip to your local pharmacy might be more convenient than one to your vet clinic. Of course, reduced cost can be the most "valuable" benefit; drugstores are often able to obtain large-volume discounts, which are passed along to the customer.

However, problems can arise when the details of the prescription are transmitted from one place to another—in this case, from your vet's office to the drugstore. On one end the prescribing veterinarian or clinic staff member calls or faxes in the prescription, and on the receiving end await a drugstore staff member (or maybe two) and the resident pharmacist. Just as in the popular children's game "Gossip," the more people involved in the circle of communication, the more likely it is miscommunication will occur.

Online

When I first became a veterinarian, I never dreamed my clients would be able to fill prescriptions online. However, online pharmacies' willingness to compound medication, their ability to provide discounted prices, and today's preference for "home shopping" with a mouse and credit card has made this a legitimate choice in many people's minds. Some online pharmacies are set up strictly for veterinary prescriptions, some for human prescriptions, and others for both.

Many veterinarians have strong reservations about the online option. They are concerned about a lack of quality control, as well as their inability to double-check the medication their patients are receiving. Not all online pharmacies subscribe to the same school of ethics. Some sell medication without a valid prescription or willingly provide more than the prescribed amount. At the time of publication, regulation of online sources is questionable. If you feel an online pharmacy is the best way to fill your dog's prescriptions, ask your veterinarian for her recommendations and adhere to them.

What Should Appear on Your Dog's Rx Label?

- Date the medication was prescribed
- Your dog's name
- Your last name
- Name, address, and telephone number of the prescribing veterinarian
- Amount of medication (milliliters, ounces, number of pills or capsules)
- Strength of the medication
- Dosage and duration
- Route of administration (i.e., orally, applied to the skin, in the ear)
- Number of refills
- Cautionary instructions ("Shake well," "Keep refrigerated," "Don't let your dog drive")
- Expiration date

- **How long should the medication be administered?**

Find out whether it is for a set number of days, until it runs out, or indefinitely. Don't assume that just because the pill vial is empty, your vet wants it discontinued.

- **What happens if a dosage is accidentally skipped? (We're only human, after all.)**

Perhaps your vet will want you to double-up the next dose, or it may be okay to just go ahead and skip one. Be sure to ask ahead of time.

- **Do I need to worry about handling the medication?**

Double-check that there are no issues with you handling the drug, especially if you are pregnant.

Questions about Special Diagnostic and Therapeutic Procedures

These days, we veterinarians have all kinds of nifty ways to diagnose and treat our patients that, in the past, could only be accomplished surgically. As I described earlier in this book, there are MRI scans, CT scans, ultrasound guided biopsy procedures, endoscopy, fluoroscopy, nuclear medicine studies, and the list goes on and on. These techniques are fabulous because they are far less invasive than surgery, though they can still pose some risk to the patient. Satisfactory answers to several questions are essential before giving your vet the "green light" to proceed with these tests.

- **How will this procedure be of benefit?**

Before saying, "Yes," make sure the logic behind the procedure makes sense. Will the results of the diagnostic procedure have the potential to change what happens next and provide some necessary peace of mind? If the procedure is meant to be therapeutic, what is the likelihood of success for your dog? Will it be the be-all and end-all, or will more steps be recommended?

- **Does the procedure require general anesthesia?**

For many of these tests, general anesthesia is required for the safety and comfort of the patient, or because the dog must be completely immobilized. Sometimes sedation is adequate. Reference the questions about general anesthesia, p. 148.

- **What are the risks?**

For some noninvasive procedures, such as MRI or CT scans, the only significant risk is the general anesthesia. For others, such as an ultrasound guided biopsy, there is a chance of bleeding or inadvertent tissue trauma. Find out what can be done to minimize such complications. You want to feel confident that the benefits of the procedure outweigh the risks before proceeding.

- **Are there any special instructions prior to the procedure?**

You may be asked to withhold food and water, discontinue some medications and start others. If it's an abdominal ultrasound, your vet might ask you to bring your pup in with a full bladder and empty stomach (if you can!)

- **How long will my dog be in the hospital?**

Many of these tests involve just dropping your dog off in the morning and picking him up at the end of the day. If your dog is ill or has other complications, he may need to stay in the hospital for a longer period.

- **What can I expect during the recovery process?**

Your dog may be able to resume his normal routine right away, or there may be a recommended period of confinement or other special instructions. It's always helpful to know this in advance, so you can orchestrate your life appropriately.

- **How many times have you performed this procedure?**

Breathe a sigh of relief if you learn that your veterinarian is experienced. But if that's not the case, strongly consider asking for someone with greater expertise.

- **How often do complications occur?**

Yet another tricky, but essential question—I usually answer this in the form of a percentage, and then qualify my answer a bit if I think that the dog we're talking about carries a greater than normal risk.

- **Will pain management be necessary?**

Some special procedures create not even a smidge of pain or discomfort. Others (the ones that involve needles and biopsies) are definitely capable of hurting. Find out which, if any, pain medications your vet has in store for your pup, and their side effects (see p. 140).

Questions about General Anesthesia

People often express considerable concern when it comes to anesthetizing their dog. Some risk is always there, but when a dog is appropriately prescreened and monitored carefully during general anesthesia, complications are few and far between.

When your vet discusses using "chemical restraint" of some fashion, pay close attention to her terminology. In general, *sedation* refers to *tranquilization*—the patient is still somewhat awake, responsive, and capable of experiencing pain. *Anesthesia* implies complete loss of consciousness. Some veterinarians use the terms incorrectly—they might talk to you about "sedating" your dog when he will really be anesthetized. The distinction is important, because the risks and complications associated with anesthesia are far greater than with sedation. So, when your vet says, "sedation," by all means have her clarify the term.

- **Is it truly necessary to anesthetize my dog?**

Some procedures absolutely require general anesthesia, and in most cases it's inhumane to do anything less. And with other tests movement of any kind (other than breathing, of course) can jeopardize their success. There are times when general anesthesia might be considered optional—for example, removal of a small skin mass might be readily accomplished using sedation in conjunction with a local anesthetic.

- **How can we be certain my dog is a good candidate for anesthesia?**

Basic screening, including a thorough physical examination, and blood and urine testing is an important part of assessing your pup's anesthetic risk. Abnormalities might direct your vet toward a more advanced evaluation before your dog is anesthetized or use of different anesthetic drugs.

- **Will an intravenous (IV) catheter be placed prior to anesthesia?**

Even if the anesthetic procedure is short, the last thing anyone wants is a significant complication to arise without there being *venous access* (an intravenous catheter in place). Complications of the anesthesia itself, or the operation for which the patient was anesthetized (for example, drop in blood pressure, blood loss, heart rhythm abnormalities) may well require intravenous delivery of medication and fluids. Since placing an IV catheter can be tricky and time-consuming, doing it beforehand is, in my opinion, a must.

- **What type of anesthesia will be used?**

Anesthesia can be administered by way of injection or via inhalation of a gas. If the procedure is quite short, often injectable anesthetic drugs alone are used; with lengthier procedures, it is common to induce anesthesia with these drugs, then maintain the anesthetic state with gas.

- **Will an endotracheal tube be used?**

This is a tube that is placed into the windpipe after a dog has been anesthetized (the process is called *intubation*). Oxygen alone, or oxygen and anesthetic gas, can then be provided via this tube. This can also be accomplished, less effectively, by using a mask placed over the dog's nose and muzzle. The endotracheal tube provides some definite advantages, including more accurate delivery of oxygen and anesthetic gas; increased ability to actively assist breathing (this is critical if an emergency arises); and prevention of aspiration pneumonia (inhalation of vomited material into the lungs). Whenever a dog is under general anesthesia, having an endotracheal tube in place is a good insurance policy.

- **Who will induce and monitor anesthesia?**

In addition to the veterinarian, registered veterinary technicians (RVTs) are allowed to place a patient under anesthesia (but only under the supervision of a vet). Ideally, a technician (though it doesn't have to be an RVT) or another veterinarian should continuously monitor your dog's anesthesia.

- **What anesthetic monitoring equipment will be used?**

Today, sophisticated electronic equipment allows the veterinarian and her staff to monitor a number of physiological parameters, including heart and respiratory rates, blood pressure, an electrocardiogram (ECG), blood-oxygen saturation (*pulse oximetry*), and body temperature. Such monitoring is used to detect problems early before they've had a chance to cause significant complications. Not all of these parameters need to be monitored in every anesthetized dog. In general, the longer the duration of anesthesia and the riskier the procedure, the more such monitoring devices are recommended.

- **Where will my dog recover?**

The answer you want to hear is, "In our treatment room (or intensive care unit) where your dog will be monitored 100 percent of the time until he has completely recovered from anesthesia." Any answer that does not involve continuous monitoring is the wrong answer.

- **What complications might arise?**

Some risk is associated with general anesthesia even with the youngest and healthiest of patients, though anesthetic risk tends to increase with age and illness. A dog with kidney or liver failure has a decreased capacity to metabolize the anesthetic drugs, and one with heart disease may have difficulty maintaining normal blood pressure. B*rachycephalic* dogs (Boston Terriers, Pugs, Pekingese—the ones I affectionately term "smoosh-faced") are predisposed to breathing difficulties when recovering from anesthesia. Talk to your vet about any potential anesthetic complications unique to your dog.

Questions about Surgery

No matter the type of surgery, there are a number of things to know before letting your dog set "paw" into the operating room. Remember, there is no such thing as "routine" surgery, and even a commonplace procedure, such as neutering (see p. 156), can result in serious, life-threatening complications.

There are two questions that clients frequently ask. The first is, "How long will the incision be?" I chuckle at this, because it always reminds me of when people ask each other, "How many stitches did you get?" Know that incisions heal side-to-side, and not end-to-end. So, an incision that is 6 inches long will heal just as quickly as one of 2 inches! (Actually, a tiny incision always worries me, because I wonder how in the world the surgeon could have done a thorough and safe job through such a tiny opening.) The answer to this popular question is always, "The incision will be as long as the surgeon needs it to be to do the job well."

59% of dog owners think their furry companions are good for their health and will help them live a longer life.

Question two is, "How much hair will you have to shave?" Unless you live in a frigid climate, or your dog is scheduled for a show in the very near future, a temporary bald spot, however significant, is a small price to pay for a thoroughly sterile surgical site. Besides, your dog could care less about the answer to this question—and his "clip job" will be a great conversation-starter at the dog park.

Now, here are some less commonly asked, but far more important questions about a prospective surgical procedure:

- **Why does my dog need surgery?**

What you really want to know is what are the *alternatives* to surgery and what will happen if he doesn't have it. The risks and benefits of any nonsurgical options should be discussed and compared to those of the recommended surgical procedure. You want to feel extremely confident that surgery is truly the best option.

- **Is there more than one way to do the surgery?**

It's not unusual to find more than one surgical technique available to accomplish the task at hand. For example, when a dog tears his cruciate ligament (a ligament in the knee joint), there are a few different surgical techniques to choose from. Some are much more "state of the art" than others. If your vet doesn't perform one of the more advanced techniques, consider consulting a board certified veterinary surgeon. Remember, second opinions are always okay (see chapter 9, p. 207).

- **What are the potential complications?**

Not only are there the universal risks of anesthesia (see p. 248) and surgery to talk about, your dog may have his own risks based on his age, overall condition, and other medical issues.

Risks should also be evaluated with consideration of the lifestyle you lead and conditions at home. For example, do you have stairs that would prevent you from adequately confining your dog after orthopedic surgery, so that the bones can heal properly? Will you be home to give him his antibiotics and prevent him from chewing at his stitches?

Some veterinarians are loathe to describe potential problems for fear they will sound like the disclaimer at the end of a prescription drug commercial. Let your vet know that you really wish to hear it all!

- **What are the pre-operative instructions for this surgery?**

Find out the particulars of withholding food, water, and exercise prior to the procedure. Ask if any medications or supplements are to be started (antibiotics, pain medication, vitamins) or discontinued (nonsteroidal anti-inflammatory medication can promote bleeding) prior to surgery.

- **How long will my dog be in the hospital?**

This is important for a couple of reasons: your nerves and your bank account. Typically, the longer your dog stays in, the more costly it is for you financially and possibly emotionally!

- **How is my dog to be cared for through the night?**

Following surgery your dog needs 24-hour care and supervision. If the facility where surgery was performed does not offer overnight care, find out if you will be allowed to take your dog home for the night or transport him to an after-hours emergency-care hospital.

- **What can I expect the recovery process to be like?**

This is very important to know *before* surgery, especially when you have commitments in your life besides the canine center-of-your-universe. You want to have everything prepared so you can follow instructions properly when your pup comes home from the hospital. If you work, you might need to arrange for time off, and if confinement is necessary, you may need to set up a "doggie playpen" before the patient arrives home. When your dog comes home with the dreaded Elizabethan collar—the notorious plastic neckwear that looks like a satellite dish—you need to get busy storing all of your cherished breakables.

- **How many times have you performed this surgery?**

A wonderfully reassuring answer sounds like, "I've performed this procedure 10 to 20 times a year for the past five years." Less comforting is, "I've done this surgery a couple of times." No matter how much you love your vet, you want experienced hands; yes, every vet needs to learn somehow and somewhere, but not without close supervision (as in a teaching hospital).

Some procedures are performed day in, and day out, in general practice settings. Others tend to be performed only by board certified surgical specialists. Then, there is everything in between—the operations that one family vet will perform while another feels better referring to a specialist. It's these in-between cases that take some figuring out. I always laugh when a client asks *me* to do a recommended surgical procedure; as a practicing internist, I'm about as far removed from surgery as you can get.

- **How often do complications occur?**

I know that it takes some courage to ask this question of your vet—it sounds confrontational, but trust me, a veterinarian will respect you for asking. Yours should be able to give you a rough idea of the likelihood of complications. She may qualify this percentage a bit based on risk factors unique to your dog.

- **How will my dog be assessed and managed for pain?**

Pain management for dogs has come a long way over the past couple of decades. Veterinarians and their staff are much more educated these days about evaluating pain and are trained to deliver new and improved medication. Injectable drugs are typically started before or during surgery rather than just following the procedure. As you know for yourself, preventing pain at the start consistently provides better results than trying to get a handle on the pain later. A number of effective oral pain medications can be prescribed for when the patient goes home.

Questions about Spaying and Neutering

Most dogs are spayed or neutered at a young age when they are robust and healthy. Because of this, and the acceptance of neutering as a "routine procedure," it's easy to be lulled into a sense of complacency. But, oh no—there shall be no rest for the medical advocate you are striving to be! Please ask all of the universal questions listed on pp. 135–139 pertaining to surgery and anesthesia. They should be answered to your satisfaction before saying "Yes" to spay or neuter surgery for your dog.

- **How old should my dog be when surgery is performed?**

Some people believe in very early spaying and neutering, beginning anytime after six weeks of age. Surgery at this young age is used primarily as a means of ensuring that all shelter animals are neutered before adoption. Outside the shelter situation, dogs are more commonly neutered after four months of age. Allowing more time for some growth and development tends to create a less risky anesthetic and surgical procedure.

For female dogs, the time to spay is before she experiences her first heat, which typically occurs between six and 12 months (in general, the larger the dog, the later the first heat cycle). Why before the first heat cycle? There are a couple of reasons. Spaying before then almost completely eliminates the possibility of breast cancer later in life. And, the last thing you want is an unplanned pregnancy for

your dog. People are surprised by the power of canine hormones to attract males—even from great distances. Drop your guard for one second, and you have a pregnant girl on your hands.

A male should be castrated in a timely enough fashion (before he is 12 months) to thwart undesirable "bad boy" behavior, such as aggression toward people or other dogs; running away from home to follow the scent of a female in heat; or urine-marking in inappropriate places, such as the corner of the living room sofa. If the goal of neutering is to prevent testicular cancer and some types of prostate gland disease (unfortunately, neutering doesn't protect against prostate gland cancer), your dog should be neutered before he's three years old.

If you have a purebred dog you are considering using for breeding, the decision about whether or not to neuter may not be made until later—but in order to maximize the health benefits of the procedure, I recommend the choice be made by the time he or she is two.

- **Should any other procedures be performed at the same time?**

Take advantage of the general anesthesia required for spaying or neutering to accomplish any other needs, such as removing dewclaws; repairing an umbilical hernia; or removing baby teeth that never managed to fall out on their own. It's always a good idea to avoid more anesthetic procedures than necessary.

Questions about Breeding Your Dog

When you contemplate breeding, my hope is you have a valuable, healthy purebred dog on your hands—one with an excellent disposi-

tion who has undergone appropriate screening for inherited diseases (see chapter 1, p. 16, for details). The same criteria apply to your dog's intended mate. Additionally, when your dog is the female of the pair, be sure you have the time and resources necessary to manage possible complications of pregnancy, whelping, and raising a handful (or two) of puppies.

- **What should be done prior to breeding?**

The male and female dogs should have a thorough physical examination within a month or so of the anticipated time of breeding. If your female dog has never been bred, your vet will want to perform a vaginal exam to check for anatomical defects. Both dogs should be current on vaccinations and deworming, as well as have a blood-screening test for *brucellosis*, a venereal disease that can cause serious illness in the dam, sire, and pups. Your vet may have other pre-breeding recommendations, based on the breed, health issues, and prior breeding history.

- **How should ovulation timing be evaluated?**

Knowing when the egg is released from its follicle on the ovary and is mature provides a way of maximizing fertility. This ovulation timing also takes away the guesswork of knowing exactly when pups will arrive. It is especially important when access to the male dog or the amount of semen for artificial insemination is limited. Your veterinarian will outline a plan for ovulation timing using blood tests to measure hormone levels (*progesterone* and *luteinizing hormone*), vaginal *cytology* (looking at vaginal cells under the microscope), and *vaginoscopy* (the use of a viewing scope to evaluate the tissue folds within the vagina).

40% of people report that owning a dog motivates them to exercise on a regular basis.

- **Which breeding option is best?**

Given the fact that intended mates often live hundreds, if not thousands of miles apart, it's a good thing there are options other than natural breeding. Artificial insemination has become increasingly popular because it offers a number of benefits compared to natural breeding, and when managed properly, produces excellent conception rates. It's far easier to ship frozen semen (or even freshly collected, chilled, extended semen) than it is to ship an entire dog from one coast or continent to another! Other advantages: it provides the opportunity to evaluate the stud dog's semen quality; semen from a valuable male can be frozen and used for artificial insemination to produce offspring long after he is gone; and finally, artificial insemination may be the only way to obtain success when the male or female lack "experience."

Success of any artificial insemination technique directly correlates with the experience of the person performing the procedure. Semen can be deposited into the vagina or directly into the uterus. Both surgical and nonsurgical techniques are available for intrauterine insemination (neither has been shown to provide superior results); the nonsurgical method does not require anesthesia, so is generally preferred. Consult your vet about which insemination technique makes the most sense for your situation.

- **How should my dog's pregnancy be monitored?**

Besides observing the pregnant dog carefully at home, it may be recommended she has an abdominal ultrasound evaluation early on (at 28 to 30 days) to confirm pregnancy. Ultrasound can also be used throughout to monitor the health of the fetuses. It's a good idea for your veterinarian to examine your pregnant dog a couple of weeks prior to the expected delivery date to ensure that everything is as it should be. This is also a good time to assess the litter size (best accomplished with an X-ray), and discuss your role in the whelping process (see p. 161). Consider the Whelpwise System™ for at-home monitoring of uterine contractions and fetal heart rates, which makes it clear when labor has started and whether or not it is progressing normally (www.whelpwise.com).

- **What should I feed my dog?**

Ask your veterinarian about specific dietary changes during pregnancy as well as when nursing. In general, a diet richer in calories and protein is recommended, but the type and amount of food varies based on breed, prior pregnancy history, and overall body condition. Some female dogs are prone to *eclampsia*, a condition in which their body is drained of adequate calcium levels during the nursing process. Your veterinarian will be able to offer ways of nutritionally avoiding this potential problem. Do not give your pregnant or nursing dog any dietary supplements without first consulting your vet.

- **What should I watch for during labor and delivery?**

The answer to this is of paramount importance, because recognizing the abnormal is the key to preventing whelping complications for mother and puppies. You need to know what "normal" looks like in terms of the stages of labor; "mom's" actions during and immediately following whelping; the newborn puppies' behavior; vaginal discharge; and appearance of the afterbirth. Become familiar with normal time intervals between onset of contractions and appearance of the first puppy; time intervals in between puppies; and the maximum time a puppy should be visible in the birth canal before delivery. Ask your vet for places you can find normal whelping parameters and guidelines. And, I encourage you to be an observer a few times before supervising a whelping process on your own.

- **What am I empowered to do at home versus what should prompt a hospital visit?**

Most vets recommend that they be the ones to pull a fetus stuck in the birth canal or give drugs to stimulate contractions. It is up to you to work out a plan with your vet (well in advance of the due date) that feels right to everyone involved and is in the very best interest of mother and her pups.

- **What should I do with the puppies after they are born?**

Close supervision without excessive interference is the best rule of thumb. Talk to your vet about the parameters you should monitor. Is the mother stimulating the newborns to breathe? Are the puppies vocalizing and nursing effectively? With your vet's counsel, you can be properly prepared to help perform tasks such as breaking the sac covering a puppy, correctly tying and disinfecting the umbilical cord, and clearing the airway and stimulating breathing if needed.

- **Who should be called in case of emergency?**

It pays to have all your ducks in a row on this one. The last thing you want is to make a bunch of phone calls to figure out who can help you while in the midst of a whelping emergency.

- **How soon after whelping should mother and pups be evaluated?**

Check to see how soon your vet wants to examine the entire gang. She will inspect the pups for any birth defects; evaluate "mom" for milk production, mastitis (infection of the mammary gland), and any birthing complications; and possibly give an injection of *oxytocin*—a drug to help "clean out" the uterus.

SECRET FOR SUCCESS

The Importance of Following Your Vet's Instructions

Ever since there have been veterinarians, they've been making jokes about client compliance, or lack thereof. The joke is, if we want a medication to be given to a dog twice a day, we write the prescription for three times a day. If we want to control obesity by feeding the dog only three cups of dog food a day, we recommend two cups.

All kidding aside, the reality is most of our clients cannot, or will not comply with all that is asked of them. No one likes administering medical treatment to their dog at home. Few dogs enjoy the process, and people who are crazy about their dogs try to avoid doing things that their dogs aren't crazy about.

Persuasive Pups

Dogs have an uncanny knack for "talking" their unwilling, inexperienced humans out of getting the job done. I've witnessed clever dogs who can work their tongues around peanut butter, cream cheese, or filet mignon, swallow the edible camouflage, then spit the "hidden" pill back out as though it were a watermelon seed. Other dogs defy compliance with a growl, a show of teeth, or a pair of pitifully sad brown eyes.

Time Constraints

There's also the matter of busyness. Most of us are far too pressed for time as it is—so who's supposed to give that pill in the middle of the day when you are at work? Does your vet really expect you to cancel your trip to Mexico in order to bring your pup in for a recheck visit in three weeks? How are you supposed to feed one dog a special diet, when all four of your dogs nibble from the same food bowl throughout the day? It's easy to see why corners get cut a bit in the compliance department. Heck, I've had to treat my home menagerie for a variety of maladies, and I have difficulty complying with my own recommendations!

- **Should we consider a "planned" Cesarean section?**

If your dog had whelping problems before, or if she is a breed notorious for needing a C-section—such as a Bulldog, Chihuahua, or Boston Terrier—talk with your vet about the pros and cons of having one. A planned C-section is always preferable to middle of the night or weekend emergency surgery.

Get Creative

Believe it or not, when your veterinarian gives you a laundry list of treatment instructions, she isn't purposefully trying to make your life crazier. Just like you, her primary focus is your dog's health. In fact, she may be so wrapped up in thinking about your dog that she forgets you have to work for a living. I encourage you to be up front with your concerns so that you and your veterinarian can come up with creative solutions.

Perhaps the timing of recommended treatments can be tweaked, or an antibiotic can be chosen that only needs to be given twice rather than three times a day. Maybe the hospital offers weekend appointments and the timing of the recheck visit can be adjusted so that you won't have to cancel your trip to Mexico. If your dog is too clever to take his pill slathered in pâté, ask the staff for advice. They usually have a number of tricks up their sleeve and are experienced at outsmarting even the cleverest of dogs.

Understand Instructions

Should the hot compress be applied over the wound or around the wound? How hot should it be? What if your precious pooch tries to bite (perish the thought) when you attempt to apply the compress? Ask all the questions necessary to achieve a thorough understanding of home treatment, and if you've been asked to do more than one thing, request written instructions. If these aren't clear or do not work, call the vet staff for help. I provide a treatment sheet on my website (www.speakingforspot.com). It's the form we use in my hospital to help keep track of treatment plans.

- **How should the puppies be monitored and cared for?**

Should the pups be weighed on a daily basis? How soon should "mom" be allowed (should she choose) to spend significant time away from them? What are the appropriate ages for introducing dog food, deworming medication, vaccinations, weaning, and going to a new home?

"I describe 'cancer' as the one word in the entire English language that the mind sees in all capital letters."

—VICKIE GIRARD,
AUTHOR OF *THERE'S NO PLACE LIKE HOPE: A GUIDE TO BEATING CANCER IN MIND-SIZED BITES*

8

When the Diagnosis Is Cancer

Cancer, neoplasia, growth, tumor, malignancy, the big "C": no matter which word is used, it is the diagnosis we all dread. It's not that cancer is always associated with a terrible outcome—not by any stretch of the imagination. It is true, however, that whenever cancer is diagnosed, it is inevitable that lives are going to change. And change isn't something we necessarily relish when it comes to our canine family members.

Accompanying a beloved dog through the diagnosis and treatment of cancer can be an emotional roller coaster ride.

Perhaps you are wondering why, out of the many diseases discussed in this book, I've devoted an entire chapter to cancer. I had a couple of reasons in mind when I sat down to write. Cancer has become a very common disease that afflicts far too many of our beloved pets, and there's little doubt that its incidence is on the rise. We know that cancer occurs more commonly in older dogs, and we're doing a good job keeping our dogs living and thriving longer than ever before. Plus, our abilities to detect cancer are continually improving. It is estimated that one in three of our dogs will be diagnosed with it at some point during their life. That's a whole lot of cancer we dog-lovers will be facing.

ANDREW I'll never forget my patient Andrew, and the two-and-a-half years I had the good fortune of taking care of him. He wasn't a winner at the racetrack, but a greyhound rescue volunteer named Jill instantly recognized another kind of "winner" when she looked into the eyes of this gentle giant. She signed his adoption papers and Andrew came to live with her and her husband and baby daughter. It's a good thing Jill had finely honed medical advocacy skills because Andrew was going to put them to the test over and over again.

Jill's first decision-making challenge occurred shortly after Andrew's ninth birthday when he was diagnosed with lymphoma—a cancer involving white blood cells called **lymphocytes**. As soon as Jill learned of the diagnosis, she found out all she could about lymphoma, and the potential risks and benefits of chemotherapy. She found out that even with successful chemotherapy, the long-term outlook is poor—an average survival time of 12 months. After much investigation and contemplation, Jill opted for chemotherapy, but advised me she was committing to only one treatment. After that she would decide whether or not it was reasonable to proceed with more.

We treated Andrew and, within a week, all of his enlarged lymph nodes disappeared. He did not experience vomiting, diarrhea, or loss of appetite. In fact, Andrew was feeling better than he had in a long time. Jill was thrilled, and she opted to proceed with an extended protocol utilizing a number of different drugs (**combination chemotherapy**), always with the stipulation that at her request, we could stop at any time. Fortunately, it was smooth sailing for Andrew. Chemotherapy was providing the very best possible outcome. Andrew was living his life to the fullest and Jill and her family relished all the time spent with their big dog.

Eight months into chemotherapy, disaster struck. One of Andrew's intravenous drugs managed to seep out into the tissue surrounding the vein, where it created a very painful inflammatory reaction. The effect on the tissue was so severe that surgery and multiple bandage changes were deemed necessary.

For Jill, this second go-around of decision-making was far tougher than the first. Was it reasonable to ask Andrew to endure treatment for this devastating chemotherapy complication, knowing that his lymphoma could recur at any time? Jill had already experienced an incredible eight-month "honeymoon" with him as a result of his treatment. Should she be satisfied with that, or should they take on this new challenge? After a great deal of soul-searching and several nose-to-nose discussions with Andrew, Jill agreed to treatment again with the understanding that, at any point, she could call it a day, if that's what she sensed her beloved dog wanted.

Andrew rallied in response to his therapy, and within a few weeks, his leg improved dramatically. But now Jill was faced with yet another major decision. After all she and Andrew had been through, how did she feel about the notion of continuing chemotherapy? Jill had to dig deep to come up with the answer. Knowing it was impossible to predict how long the lymphoma would stay away, with or without more treatment, Jill came to a conclusion based on two factors:

First and foremost, she considered how Andrew might feel about continuing to come to the veterinary hospital after all he'd been through. The fact of the matter was, Andrew never seemed to mind his trips to see us at all. I think he rather enjoyed all the attention (and the cookies). Second, Jill considered which of the two options—ongoing chemotherapy versus discontinuation of treatment—would in the long run best serve her own peace of mind (see more on this subject on p. 32).

Jill opted to proceed with chemotherapy as long as any drugs that could cause harm if injected outside of the vein were excluded from the treatment protocol. This produced excellent results. For Andrew's one-year, post-diagnosis anniversary, we had a celebration complete with cake ("dog-edible," of course) and balloons. We did the same for his two-year anniversary.

Midway through his eleventh year, Andrew began limping. X-rays revealed some horrible news: he had osteosarcoma in his front leg. This was bone cancer completely unrelated to the lymphoma. Now Jill had to decide whether to have the affected limb surgically amputated or to provide Andrew with pain medication until it was evident his pain was no longer manageable. After again doing her research and "talking things over" with her dear dog, Jill opted for palliative care rather than surgery.

Pain medication carved out a valuable period of closure for the two, during which time they did all of their favorite things together. And, when it was finally time for Jill to bid her beloved Andrew goodbye, I administered the injection of euthanasia solution while he was encircled in Jill's embrace.

I'd watched this glorious dog evolve from a lithe, middle-aged running machine into a heftier, gray-muzzled senior citizen. I'd watched Jill's daughter transform from a shy toddler dwarfed by her big dog to a talkative youngster whose head almost reached the top of his shoulders. And, I so admired Jill for the difficult decisions she consistently met head on.

When I spoke with Jill a few weeks after Andrew's death, we talked about all that she and her dog had been through. She told me that every decision had been difficult, but she carried no regrets. They had enjoyed two-and-a-half extra years of quality time together. I told Jill that, throughout the entire experience, I believed she had been a remarkable medical advocate for her wonderful four-legged friend.

There's a second reason I believe that cancer deserves its very own chapter. More than any other diagnosis, cancer requires us to make uniquely challenging and emotionally taxing medical decisions. Accompanying a beloved dog through the diagnosis and treatment of cancer can be an emotional roller coaster ride, capable of clouding judgment and stirring up tremendous uncertainty and angst. My goal is to provide you with some practical guidance and emotional fortitude, if and when cancer challenges your dog's health.

16 million Americans say they are as attached to their pets as they are their best friends.

What Causes Cancer?

When it comes to a cancer diagnosis one of the most common questions I'm asked is, "How did my dog get this?" Unfortunately, it's exceedingly rare that I am able to provide a clear-cut answer. Yes, we know that cigarette smoke, asbestos, sun exposure, and some pesticides and lawn herbicides can be carcinogenic in dogs. We also know that female hormones influence the development of mammary tumors (breast cancer). In most cancer cases, however, there is no discernible cause.

Cancer happens when cells manage to elude the body's normal growth-control mechanisms, then multiply in a disorderly, uncontrolled fashion. It is believed that certain individuals are inherently predisposed to developing such renegade cell populations—we certainly see an abundance of cancer in particular breeds, including

QUICK REFERENCE

Types of Cancers

Carcinomas: Cancers Arising from Cells that Line Body Tissue

Name of Cancer	Tissue Origin
Anal sac adenocarcinoma	Anal sac
Mammary gland adenocarcinoma	Breast tissue
Prostate gland adenocarcinoma	Prostate gland
Pulmonary carcinoma	Lung
Squamous cell carcinoma	Oral cavity and skin
Transitional cell carcinoma	Lower urinary tract (bladder and urethra)

Sarcomas: Cancers Arising from Connective Tissue

Name of Cancer	Tissue Origin
Chondrosarcoma	Cartilage
Fibrosarcoma	Fibrous connective tissue
Hemangiopericytoma	Small blood vessels
Hemangiosarcoma	Blood vessels
Histiocytic sarcoma	Macrophages (immune system cells)
Leiomyosarcoma	Smooth muscle
Liposarcoma	Fat
Lymphoma (lymphosarcoma)	Lymphocytes (type of white blood cell)
Mast cell cancer	Mast cells
Melanoma	Melanocytes (pigment producing cells)
Osteosarcoma	Bone

Leukemias: Cancers Arising in the Bone Marrow

Name of Cancer	Tissue Origin
Lymphocytic leukemia	Lymphocytes (type of white blood cell)
Myelogenous leukemia	Myeloid cells (type of white blood cell)
Multiple myeloma	Plasma cells

Boxers, Golden Retrievers, Rottweilers, Bernese Mountain Dogs, Boston Terriers, English Bulldogs, Scottish Terriers, and Cocker Spaniels. Giant dog breeds (heavier than 75 pounds) are predisposed to bone cancer. Nasal cancer is most commonly diagnosed in *dolicephalic* breeds ("long-nosed" breeds, such as German Shepherds).

In addition to inherent predisposition, some sort of trigger (environmental factors, hormones, an aging immune system) is likely necessary for the cancer cells to wreak their havoc. Unfortunately, for dogs with cancer, the underlying cause is rarely obvious. Hopefully, more complete answers to the question, "How did my dog get this?" will be forthcoming in the not too distant future.

Cancer Lingo

A number of different terms are used when discussing canine lumps and bumps, growths and nodules, masses and tumors, malignancies and cancers. And guess what? Two different people (even two veterinarians) may have entirely different notions about what exactly some of these words mean. It pays to listen carefully—far better to ask for clarification than to assume that the meaning of the word your vet is using is the same as the meaning with which you may be familiar. Here is a "starter list" of definitions for some common cancer lingo. I've offered the terms in the order you are likely to confront them—from the beginning stages of cancer diagnosis through treatment.

Tumor, mass, neoplasm: all three terms are synonymous. They refer to an abnormal proliferation of cells that develops in spite of regulatory systems in the body designed to keep this cellular replication in check. Tumors, masses, and neoplasms are umbrella terms in the sense that they can be benign or malignant (see below).

Biopsy: the collection and microscopic examination of a solid tissue sample.

Cytology: the examination of cells or fluid under the microscope. Such samples are typically obtained via *fine needle aspiration*, in which a small needle (no bigger than the size of a vaccination needle) is introduced into the abnormal tissue. Suction is applied via the syringe to obtain fluid or a smattering of cells.

Benign versus malignant: *Benign* tumors are good guys because they don't invade the surrounding structures or spread to other sites in the body. Although proliferating in an abnormal fashion, benign cells typically have a normal appearance. A *malignant* tumor is also an unregulated proliferation of cells; however, malignant cells usually have an atypical or bizarre appearance. They often invade into surrounding structures, interfere with normal organ function, and spread to other sites in the body via blood or lymphatic vessels. A benign tumor generally offers the better prognosis, but this isn't a guarantee. For example, a dog with a very small, very early malignant breast tumor that hasn't spread (metastasized—see p. 172) and is treated surgically might outlive a dog with a benign liver mass that is too large to be surgically removed.

Cancer: a tumor that is malignant.

Primary tumor: the original tumor as opposed to one that has spread from another site.

Metastasis: the spread of cancer from the primary site to secondary sites elsewhere in the body. This occurs when cancer cells enter the blood stream or lymphatic system. A secondary growth is referred to as a *metastatic lesion*. The term *metastasize* refers to the act of cancer spread.

Sarcomas: malignant cancers arising from the supportive tissues within the body (bone, fat, cartilage, muscle, fibrous tissue, blood, lymph). For example, osteosarcomas are tumors that arise from bone cells, and cartilage cells are the origin of chondrosarcomas (see sidebar, p. 169, for a complete list).

Carcinomas: malignant tumors that originate from epithelial cells (cells that provide the lining of various organs and glands and also comprise part of the skin). For example, a transitional cell carcinoma arises from cells that line the urinary bladder (see sidebar, p. 169, for a complete list).

Leukemia: a cancer comprised of blood cells that originate within the bone marrow, then circulate within the bloodstream (see sidebar, p. 169). When applied to leukemia, the terms *acute* and *chronic* refer to the appearance of the leukemic cells (not how long they have been present). Acute leukemias are characterized by immature blood cells, and chronic leukemias are a cancerous proliferation of fully mature blood cells.

Paraneoplastic syndrome: a disturbance caused by the cancer in parts of the body that are seemingly unrelated to the cancer site. For example, lung tumors sometimes induce a terribly painful inflammatory condition of the leg bones called *hypertrophic osteopathy*. It is unclear how the cancer causes this anatomically distant side effect, but once the lung tumor is surgically removed, the bony inflammation typically resolves.

Chemotherapy: the use of chemicals (drugs) to treat cancer.

Radiation therapy (radiotherapy): the use of a focused beam of radiation (photons, electrons, or gamma rays) to kill cancer cells.

Remission: a decrease in cancer size (also known as "tumor burden") in response to treatment. *Complete remission* means that there is no remaining evidence of the cancer. *Partial remission* means that the cancer burden has decreased in size, but remains detectable. Although complete remission is a fabulous outcome, it doesn't necessarily imply that the cancer has been cured.

Oncology: the branch of medicine that deals with the study and treatment of cancer.

Veterinary oncologist: a board certified veterinary specialist whose practice centers on the diagnosis and treatment of cancer.

Preparing to Make Decisions about Your Dog's Cancer

If your veterinarian suspects or knows that your dog has cancer, you will be asked to make a significant number of decisions. Some of them may have to do with diagnostic testing and others will pertain to treatment options. Such decisions can be tough in the best of times and, if you've just heard your dog's name and the word "cancer" used in the same sentence, the decisions can feel downright overwhelming. If you find yourself struggling emotionally, know that this is completely normal. Symptoms of grief, such as denial, anger, and depression, are common reactions to the word "cancer." Death does not hold exclusive rights to your grieving process. The really tough part is mentally reeling yourself back in, so you can do a good job addressing the matters at hand. Remember, now more than ever, that wonderful dog of yours needs you. What can you do to gain some control over the situation? Here are some suggestions.

Symptoms of grief, such as denial, anger, and depression are common reactions to the word "cancer."

- **Ask your veterinarian how urgently your decisions must be made.**

An extra day or two can make a significant difference in terms of settling down emotionally and doing the research to deal with the decisions at hand.

- **Do your best to put away preconceived, inaccurate notions of what you imagine will be your dog's cancer experience.**

Canine bodies respond differently to cancer treatment. People often get sick, experience profound fatigue, or lose their hair, but dogs rarely do. Humans may suffer depression, but dogs have the good fortune of remaining relatively emotionally unaffected, unless they react to the distress you are experiencing.

- **Read, "surf," and ask lots of questions (see chapter 9, p. 216, for details about conducting such research).**

The more you learn about your dog's cancer, the more you will feel empowered to make good decisions on his behalf.

- **Find the emotional support you need by talking with others who truly "get" what you are experiencing.**

Steer clear of those intent on convincing you that he is "just a dog," and that your money would be better invested in that new car you've had your eye on. By all means, talk about your feelings, but choose your listeners wisely. You know who among your coworkers, friends, and loved ones will understand what you are going through. Share your feelings with your veterinary clinic's staff. They may be able to connect you with another client or two who have had a similar experience. Join a support group—locally or online. It may take some looking, but there are oodles of other people out there who have emotionally "been there, done that." Whatever you do, avoid isolating yourself with your feelings and thoughts.

- **If you like to write, there's no better time to do so. Create a log or journal about your dog's illness.**

Writing may benefit you emotionally, and your entries may enhance your medical advocacy efforts by keeping track of medication, questions to ask your vet, changes in diet, contact information, Web sites to investigate, and your dog's responses to specific therapies, both good and bad. Your journal can also hold your dog's current medical record. You'll want this handy should the need for a visit to the emergency hospital arise for any reason. (Even if it is just a bee sting, the vet will want to know what type of cancer your dog has and the medication he has received.)

- **Take things one step at a time.**

Being asked to make decisions for your dog with cancer is akin to being asked to climb a tall mountain. It's strategically and psychologically better to break your ascent into small manageable increments (and there's less likelihood of tripping and falling when your eyes are not glued to the summit). Similarly, it is easier when you focus your attention on the decisions at hand rather than those that may (or may not) arise later. It is far better, at first, to decide whether or not to proceed with a diagnostic biopsy than it is to get ahead of yourself and become preoccupied with which chemotherapy protocol to administer for various types of cancer that might be diagnosed. Ask your veterinarian to prioritize the decisions before you focus on those at the top of the list.

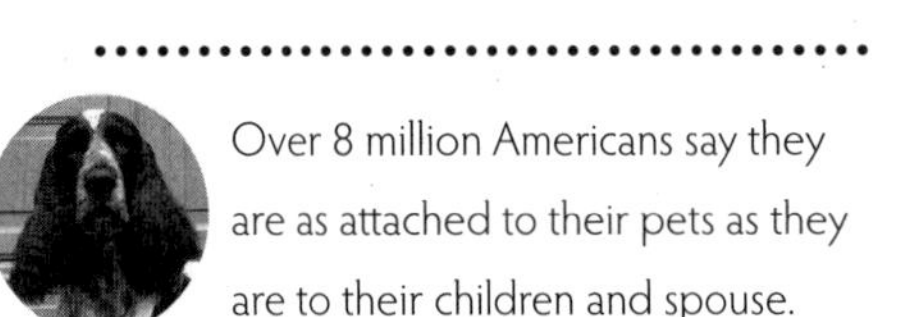

Over 8 million Americans say they are as attached to their pets as they are to their children and spouse.

The Three Levels of Decision-Making

When dealing with cancer, you will encounter three different levels of decision-making:

1. First, you'll be asked to make decisions about diagnostic tests to confirm the presence and type of cancer.

2. Next, you'll face the choice of whether or not to "stage" your dog's cancer (determine how aggressive or advanced it is).

3. Finally, there will be decisions about treating the cancer.

All three pose unique challenges. Advice for handling each level is provided on the following pages.

Level One: Diagnosing Your Dog's Cancer

People often expect that blood testing alone will be diagnostic for cancer. That's because in human medicine, specific blood proteins increase in response to the growth of certain types of cancers; for example, in men, blood PSA (prostate specific antigen) levels are a marker for prostate cancer. Unfortunately, no such tests exist (yet) in veterinary medicine. Your veterinarian may strongly suspect that your dog has cancer based on physical examination findings or abnormal test results, but in most cases, evaluation of a tissue sample (biopsy or cytology) under the microscope is needed to rule out other possibilities, such as an infectious or inflammatory disease. Such testing is also the only way to be 100 percent certain of the type of cancer. Differentiation is important because it dictates which type of treatment is likely to be of greatest benefit and allows your vet to more accurately predict response to therapy (see more on treatment on p. 186).

Microscopic Tissue Analysis (Biopsy and Cytology)

How are samples collected for microscopic analysis? *Biopsies* (collection of a core of tissue) are the most common and reliable sampling technique to establish a cancer diagnosis. In some cases, the tumor—with or without the involved organ—can be removed in its entirety (an *excisional* biopsy). Some organs, like the spleen or a kidney, are far more "removable" than others such as the liver or the stomach. If the mass cannot be fully excised because of its location or size, or its removal won't produce a better outcome, your veterinarian may recommend only a small portion of the mass be removed (an *incisional* biopsy).

The location of the mass dictates what type of surgical procedure and anesthesia are needed to collect the biopsy sample. Once the tissue sample has been surgically removed, it is placed in formalin (preservative), and submitted to a laboratory where tissue-paper-thin slices of the biopsy sample are prepared for *histopathology* (analysis of the tissue under the microscope) by a board certified veterinary pathologist.

It's a big mistake—make that a huge mistake—to go to the trouble of surgically removing a mass only to toss it out with the garbage. Gone is the chance to know the name of the cancer, whether it's benign or malignant, and whether it's been completely excised. Talk about a missed opportunity! The most common reason I hear for tossing out tissue is the vet thought the mass "looked" benign. Our eyeballs are no substitute for microscopic evaluation. If the expense of the diagnostic lab work is a problem for you at the time of the mass's removal, so be it. Just don't let your veterinarian throw the tissue away. Ask her to preserve it in a formalin container, and tuck it away somewhere safe (it will last forever). If another tumor appears, or your dog becomes sick, or you're just plain curious, you then still have access to this important information.

It's a big mistake—make that a huge mistake—to go to the trouble of surgically removing a mass only to toss it out with the garbage.

Another technique used to collect samples for microscopic evaluation is fine-needle aspiration for *cytology*. A needle no bigger than the size of one used for a vaccination is inserted into the tumor. A smattering of cells are retrieved and placed onto a glass slide that is then stained and viewed microscopically in the hopes of clarifying the diagnosis. Some vets feel comfortable interpreting these results on their own. Most prefer to submit the slides to a laboratory for evaluation by a veterinary clinical pathologist.

The advantage of the fine-needle aspirate procedure is that, because neither surgery nor anesthesia is typically required, it is far less invasive and expensive than biopsy techniques. The downside of the procedure is that, because of the small sample size collected, it results in a clear-cut diagnosis less consistently than biopsies. In general, the bigger the tissue sample, the more likely a diagnosis will be obtained. It may be beneficial to begin with the aspirate procedure in the hopes of avoiding the risks associated with collection of a biopsy specimen. Then, if the results are inconclusive, the option for collection of a biopsy remains.

QUICK REFERENCE

Tests Commonly Used to Diagnose Cancer

- Physical examination
- Blood tests (complete blood cell count (CBC), chemistry profile)
- Urinalysis
- Imaging studies
- X-rays
- Ultrasound
- CT scan
- MRI scan
- Fluoroscopy
- Fine needle aspiration
- Biopsy (tissue sample obtained with or without surgery)

Making Decisions about Diagnostic Tests

How do you decide whether to say "yes" or "no" to the tests recommended to confirm the diagnosis and type of cancer? I encourage you to proceed as follows:

Step One

Begin by asking yourself the following two questions:

❶ Are the results likely to change what happens next?

Say, for example, that your veterinarian is confident your dog has cancer, but has recommended a biopsy to determine which type is present. It makes perfect sense to consider the biopsy procedure if you are interested in pursuing treatment. If, on the other hand, you are certain you do not want to treat the cancer, regardless of type, why subject your dog (and your pocketbook) to a biopsy, when the results won't change a thing? Talk with your vet to be sure the results of the diagnostic tests have the potential to change what happens next.

❷ Will the results of the diagnostic testing provide you with some peace of mind?

As I've explained, when your pup has cancer, you've got some major decisions on your hands. Psychologically speaking, it's so much easier to proceed with decision-making when you are certain about the diagnosis—you know who the enemy is. Even if you think you do not want to treat your dog's cancer, your feelings could change when you are confronted with the decision of whether or not to put him to sleep. You might want the biopsy in order to have 100 percent certainty rather than only 95 percent certainty of the diagnosis.

An answer of "Yes" to one or both of the above questions is confirmation that the recommended diagnostic procedure is worthy of further consideration. Proceed to Step Two, below. If your answer to both questions was "No," decline diagnostics and talk with your veterinarian about other options to keep your dog as happy and comfortable as possible.

Step Two

Gather as much information as you can about the recommended diagnostic options your veterinarian has recommended and what they involve. Is anesthesia necessary? Is there risk of bleeding? What is the likelihood the results will yield a diagnosis?

What unique risk factors does your dog face? All of his concurrent health issues must be taken into consideration—for example, if your dog has severe heart disease, it might be better to obtain a fine-needle

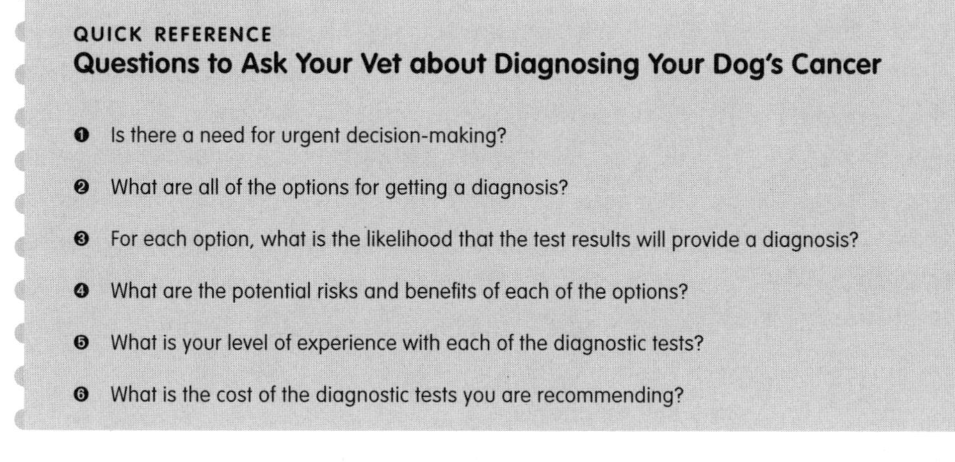
QUICK REFERENCE
Questions to Ask Your Vet about Diagnosing Your Dog's Cancer

1. Is there a need for urgent decision-making?
2. What are all of the options for getting a diagnosis?
3. For each option, what is the likelihood that the test results will provide a diagnosis?
4. What are the potential risks and benefits of each of the options?
5. What is your level of experience with each of the diagnostic tests?
6. What is the cost of the diagnostic tests you are recommending?

aspirate sample in order to avoid the anesthetic risk associated with a biopsy procedure. Create a risk/benefit analysis for possible diagnostic procedure(s). Compile a written list of all the pros and cons for comparison and go over it with your veterinarian until you feel confident that you are neither over- nor underemphasizing any significant risk factors.

If you determine that there is significant benefit and minimal risk to a procedure, give your vet the "green light." If your sense is that it is taking too big a gamble, back up and talk to your veterinarian about other options.

Level Two: Staging Your Dog's Cancer

Veterinarians have a number of ways to stage their patients' cancer. As mentioned earlier, staging (also known as grading) refers to performing diagnostic tests to determine how advanced/aggressive the cancer is. The conclusions are based on microscopic appearance as well as documentation of the various places in the body where the cancer has managed to set up housekeeping. Staging results help to predict how the cancer is likely to behave and which treatment options make the most sense.

Consider this: a dog has a small mammary (breast) tumor that could readily be removed with surgery. A savvy vet will perform a thorough physical examination, blood and urine testing, abdominal ultrasound, and chest X-rays to stage the cancer in advance of the surgery. Here is the reason why: if the breast cancer has already spread to another site, say the lungs, removal of the primary tumor would subject the dog to all of the risks of surgery without providing any real benefit.

Staging results help to predict how the cancer is likely to behave and which treatment options make the most sense.

And another example of how staging can be important: when a mast cell tumor is removed from the body, the pathologist categorizes it as grade I, grade II, or grade III, based on its microscopic appearance. The greater the number of dividing cells observed, the higher the tumor grade, and the greater the likelihood that the mast cell tumor will metastasize. After removing a mast cell tumor, if the surgeon learns from the pathologist that it was a grade I, she can rest assured that her surgery was curative. If, on the other hand, the vet learns that she removed a grade III mast cell tumor, she knows that her patient's disease was not cured surgically. She will then talk to her client about radiation therapy or chemotherapy as the next step for effectively treating the cancer. A grade II mast cell tumor is anyone's guess—some are cured with aggressive surgery alone and others tend to recur without post-op radiation or chemotherapy.

Staging also provides a thorough pretreatment baseline of cancer status and organ function to which results can be compared during and following treatment. This comparison is the most surefire way to know whether or not the cancer is truly in remission. Whether or not staging results impact treatment, some people derive tremendous peace of mind from knowing more about the prognosis and what the likely course of the disease will be.

QUICK REFERENCE
Potential Benefits of Cancer Staging

- Assessment of whether or not the cancer can be treated
- Determination of how the cancer is best treated
- Evaluation of prognosis (what is expected to happen)
- Establishment of baseline for future comparisons
- Peace of mind

Staging Tests

Staging tests usually come *after* the diagnosis has been determined, and those recommended for your dog will be based on the type of cancer that he has. Depending on cell type, many cancerous processes have rather predictable biological behavior, and they tend to spread or metastasize in a characteristic fashion. For example, prostate gland cancer tends to spread, first and foremost, to internal lymph nodes; hence abdominal ultrasound would be an important part of the staging process. Lymphosarcoma tends to set up housekeeping anywhere and everywhere in the body, so a more rigorous whole body staging is typically recommended.

- **Physical exams** are imperative. They provide information about the cancer as well as the dog's overall health. The veterinarian is looking, feeling, and listening for anything unusual. Her exam may show the location and size of the primary tumor and may document its spread to lymph nodes or other organs. The results will help direct further staging.

- **Laboratory tests** are performed on blood, urine, and a variety of fluid and tissue samples from the body. Lab test results provide information about the location of the cancer and the dog's overall health.

- **Imaging studies** such as X-rays, ultrasound, CT and MRI scans, fluoroscopy, and nuclear medicine scans are used to produce images that tell the story of what's going on inside the body. They provide a nonsurgical means for documenting the location and size of the cancer.

- **Pathology reports** provide information gleaned from the microscopic assessment of a biopsy or cytology sample (see Level One, p. 177). If surgery has been performed, the pathology report also indicates whether or not the mass was completely removed (no cancer cells extending to the tissue "margins").

- **Surgery reports** indicate what the surgeon saw during an operation. The size and appearance of the tumor as well as involvement of surrounding tissues and lymph nodes are described.

Making Decisions about Staging Your Dog's Cancer

Answers to the following questions will help you determine whether or not to proceed with staging tests for your dog.

❶ Will the results of the staging tests help determine how your dog is treated?

Amputation, the recommended treatment for osteosarcoma of the leg, would not be considered if staging chest X-rays revealed that the cancer had already spread to the lungs. Performing this staging test, therefore, makes a lot of sense. On the other hand, anesthetizing a dog with lymphoma growing in multiple lymph nodes throughout his body in order to collect bone marrow and determine whether or not the cancer resides there, too, doesn't make a lot of sense. Although the test results might impact the prognosis a bit, they would not change the recommended course of therapy.

QUICK REFERENCE

Questions to Ask Your Vet about Staging Your Dog's Cancer

❶ Is there a need for urgent decision-making?

❷ What staging tests/procedures do you recommend?

❸ What are the potential risks and benefits of each staging procedure?

❹ Which procedures will likely produce results that will determine the prognosis?

❺ Which procedures will likely produce results that will affect your treatment recommendations?

❻ What is your level of experience with the staging tests you described?

❼ What is the cost of the staging tests/procedures you recommended?

❽ What are the costs of the potential treatment options?

❷ Is your dog's overall health strong enough to withstand the staging procedures?

This is an extremely important consideration if general anesthesia is needed to collect tissue samples.

❸ Are the staging tests affordable?

It's a disservice to you and your dog to drain your budget for staging procedures that are not critical, only to find that you no longer have sufficient funds to pay for treatment of the cancer.

❹ Will the staging results provide you with some peace of mind?

Perhaps the results of the staging won't change how your dog's cancer is treated, but maybe you will sleep better at night knowing for certain the cancer is confined to just one area.

Level Three: Treating Your Dog's Cancer

Cancer is the diagnosis for which, more than any others, I wish I had a crystal ball. We vets are adept at quoting average survival times based on studies involving dozens of dogs with cancer. The truth is, we have no way of knowing what will happen to the patient sitting before us on our exam room table. It's important to remember that there will be dogs who never achieve the average survival time, and others who far surpass it. And what's really mind-boggling is sometimes the patient expected to rally, wilts, while the patient deemed to be hopeless, responds miraculously to treatment. So, how can you, your beloved dog's medical advocate, be expected to make decisions about cancer therapy?

No one ever said this was going to be easy.

I hope the following information will help you figure out whether or not you want to wage war on your dog's disease.

Treatment Options

The three most common treatment modalities (methods) used to treat canine cancer are *surgery*, *chemotherapy*, and *radiation therapy*. Sometimes, only one type of treatment is needed to effectively treat a dog's cancer—for example, a small solitary mass might be cured with surgery alone (see p. 187). It's not uncommon, however, that a combination of treatments is most beneficial, and chemotherapy or radiation therapy is used adjunctively to destroy any cells that may have been left behind after surgery (see pp. 188 and 191).

Whether or not various complementary or alternative therapies such as Chinese herbs, homeopathy, or acupuncture are capable of chasing away cancer cells is controversial. Most people, veterinarians included, do agree that they help support the cancer patient's overall health and mitigate the side effects of cancer therapy (see p. 193).

How does your veterinarian know which treatment plan is the best choice for your dog? First, she needs to know the type of cancer she is dealing with (see Level One, p. 177). Some cancer cells roll over and die in response to chemotherapy; whereas others laugh in its face. Some are sensitive to radiation therapy; others are resistant. In addition, the stage or grade of the disease allows your vet to know which treatment makes the most sense (see Level Two, p. 181).

Surgery

Surgery is ideal when the cancer exists in a single accessible site and is small enough to be removed in its entirety. Successful removal doesn't mean simply excising all of the tissue that looks or feels abnormal. Rather, the surgeon's goal is to remove all this abnormal tissue plus an additional wide "margin" of tissue that *appears* normal. Ideally, this margin should span at least 2 to 3 centimeters (approximately 1 inch) in all directions. This is done to remove any microscopic fronds of cancer cells extending from the tumor out into the surrounding tissue. The hope is that the pathologist will find "clean" rather than "dirty" margins when analyzing the tissue under the microscope after surgery.

Ask any surgeon, and she will tell you that the smaller the tumor, the better the outcome. Size influences her ability to remove the cancer and achieve clean margins. The other major surgical consideration is, as any real estate agent would say, "Location, location, location!" Removing a tumor from a dog's neck or trunk region where there is a good deal of floppy extra skin to stitch closed is far easier than removing a mass from the dog's face, leg, or tail (unless the dog happens to be a Shar-Pei or Bloodhound!) With some types of cancer, amputation of the tail or limb is recommended when it is the best option for rendering the patient free of his disease.

Sometimes surgery is performed even when it's clear beforehand that it won't result in a cancer cure. Such a procedure is called a "debulking" surgery, and its goal is to remove as much of the tumor burden as possible in order to increase the effectiveness of radiation therapy or chemotherapy. Debulking is not always appropriate; in fact some cancers (mast cell tumors, some mammary cancers) react horrifically to this partial removal. It can trigger a rapidly aggressive, inflamed-appearing, regrowth and expansion of the cancer. This is one of the reasons surgeons are happier knowing as much as possible about the cancer they're going after before making their first incision.

Chemotherapy

Chemotherapy is the use of drugs (chemical agents) to treat cancer. Each chemotherapy drug has a unique way of attacking cancer cells, and the use of *combination chemotherapy* (multiple drugs) often produces longer remission times than do single-drug protocols.

Before I go any further, I feel I must address the common misconception that chemotherapy is *always* accompanied by nasty side effects. The mere mention of the word "chemo" invariably causes an immediate and uncomfortable change in my client's expression and demeanor. Rightfully so—who hasn't had, or heard about, an awful chemotherapy experience with a friend or loved one? Please know that even though we use many of the same drugs that are given to people with cancer, dogs infrequently experience the same debilitating side effects. I discuss side effects in more detail beginning on p. 189.

Never, ever is there any obligation to proceed with chemotherapy if, anywhere along the way, you or your veterinarian aren't happy with changes you are seeing in your dog.

When you decide to treat your pup with chemotherapy, your veterinarian will create a treatment "protocol"—a recipe, so to speak, that specifies the order and timing of drug administration. The most important thing for you to know about chemotherapy protocols is that they are made to be broken. The real finesse isn't in following the protocol, but rather in knowing when and how to revise it. It is reasonable to consider postponing a treatment because of a low white blood cell count. It makes perfect sense to drop a drug from the protocol when it has caused significant side effects. Never, ever is there any obligation to proceed with chemotherapy if, anywhere along the way, you or your veterinarian aren't happy with changes you are seeing in your dog.

- *Side Effects*

Chemotherapy drugs have the greatest impact on rapidly dividing cells. Most cancers contain rapidly dividing cells—but so do the bone marrow and gastrointestinal tract. Hence, it makes sense that a low white blood cell count, loss of appetite, vomiting, and diarrhea are the most common adverse reactions to chemotherapy. The good news is that even the more common side effects tend to be the exception rather than the rule for dogs. When they do occur, they can almost always be controlled with medication and by altering the chemotherapy protocol.

White blood cells are the infection fighting cells in the body. A decreased white blood cell count can render the dog lethargic and prone to spontaneous infection. For this reason, serial blood tests to assess the white blood cell count are monitored in conjunction with chemotherapy. If the white blood cell count is found to be low, chemotherapy may be postponed or prophylactic (preventive) antibiotic therapy started.

And what about hair loss? Over time, some dogs experience hair coat changes (differences in color or texture), but significant hair loss is exceedingly rare.

Before your dog receives any chemotherapy, be sure to discuss the drug's potential side effects with your vet. And, if your dog does experience an adverse reaction, determine if any immediate action, such as a change in medication or diet, or a visit to your family clinic or emergency hospital is warranted.

- *Administration*

Most chemotherapy drugs are administered via injection into the vein (intravenous), under the skin, or into the muscle. Others are given orally. Many can be given on an outpatient basis.

Veterinary staff who administer chemotherapy should be well versed in safe handling techniques, for their own sake as well as their patients'. Drugs should be prepared under a hood apparatus that vents any fumes away from the person preparing the medication and where spills can readily be cleaned up. Personnel should wear protective masks, gowns, goggles, and gloves. It is vital the chemotherapy drugs are administered correctly and with great care. Accurate dosing is imperative and many of the intravenously administered drugs can be terribly damaging to surrounding tissue if they happen to leak out of the vein, at the very least causing your dog severe discomfort.

- *Success Rate*

Chemotherapy does not often cure cancer in dogs. Why not? Unfortunately, even when the dog is in an apparent state of complete remission, some cancer cells manage to survive in a "hidden" or "dormant" state. The analogy that always comes to mind for me is a person with a cold sore. The virus that causes cold sores persists in the body even when there's no visible trace of one. At some point, the cold sore is bound to reappear.

Not unlike the cold sore virus, cancer cells are capable of hiding in the body when all overt evidence of the cancer is gone. But they're not simply lingering, they're plotting a new offensive. They are tricky and conniving little buggers and spend their time figuring out how to outsmart the chemotherapy drugs to which they've been exposed. When the cancer does recur, it tends to be more resistant to the effects of chemotherapy. It may be possible to achieve another remission in this case (sometimes with the drugs already tried; sometimes with new drugs), but the odds are significantly poorer than during the first go-around.

Radiation Therapy (Radiotherapy)

Radiation therapy focuses a beam of photons, electrons, or gamma rays (all sources of intense energy) on the diseased organ or region of the body. The radiation is similar to that used to create an X-ray; however, the intensity is stronger and the exposure time longer.

Sometimes radiation therapy stands alone as treatment for cancer. More commonly, it is used in conjunction with surgery or chemotherapy. In most cases, radiation is used to treat a local site rather than the whole body and, as such, it provides the best results with solitary tumors rather than those that have already spread to other sites. Sometimes, radiation therapy is curative. Other times, it is administered with the hope of producing a long-lived remission. Radiation therapy also can be used as a pain management tool for dogs with bone cancer. Although the tumor isn't completely destroyed by the radiation, reduction in size may alleviate pressure, bleeding, or pain associated with the mass.

• *Side Effects*

Dogs are amazingly resilient when it comes to radiation therapy. They don't usually experience the general fatigue that people suffer. This is especially surprising, because unlike with humans, general anesthesia (albeit short-acting and light anesthesia) is required for each treatment in order to prevent the dog from moving.

Radiation affects both normal cells and cancer cells. Skin surrounding the tumor will experience acute "radiation burn," an effect that typically resembles a severe sunburn. This is usually a minor issue for the dog, though care must be taken to prevent scratching at the site. More chronic effects can include damage to the eyes, brain, or other tissue in close proximity to the radiation field. These days, machines that deliver radiation therapy are able to narrow the window of exposure dramatically, maximizing the effect on the tumor while minimizing the effects on the surrounding normal tissue.

• *Administration*

Radiation therapy is administered in two ways: with a machine that directs a beam of radiation from outside the body (*external radiation*), or using small sources of radiation (implants) placed directly in the body (*brachytherapy*). Both are intended to destroy the cancer cells.

When radiation therapy is administered with the intent to cure the cancer, it is typically divided into small fractions given on five consecutive days over the course of a few weeks. If the intended goal of radiation therapy is of a more palliative nature (reduction in size of the tumor or pain management), the treatments are administered less frequently.

Because of the size and expense of the equipment needed to generate and maintain sources of radiation, and the technical expertise required, this form of therapy is available only at private specialty or university teaching hospitals.

Complementary/Alternative Therapy

As mentioned earlier, whether complementary and alternative therapies successfully fight cancer remains a source of debate. Not nearly so controversial is the belief that complementary and alternative therapies benefit the cancer patient by supporting overall health and minimizing the side effects of traditional cancer therapy. Complementary and alternative options include Chinese herbs, dietary supplements, acupuncture, massage, chiropractic, homeopathy, and aromatherapy.

You could fill a couple of grocery bags with the number of dietary supplements, herbs, remedies, and nutraceuticals designed to support the dog and fight cancer. Many of them are untested in terms of their effects on prolonging or enhancing the life of the cancer patient, but I do find that many of my clients are truly interested in giving them with the hope of "covering all of their bases."

Most of these products are completely harmless, but I advise caution for a few reasons. First, the addition of six different new products all at once has the potential to snuff out a dog's already puny appetite. Second, some of these products have the potential to interfere with the intended action of prescription drugs or chemotherapy agents. Finally, when a new product is started simultaneously with the administration of a new chemotherapy drug, and negative side effects occur, it may be impossible to determine whether the new product or the chemotherapy drug was at fault.

Dietary Therapy

We know that diet plays a role in fighting cancer and in the prevention of *cancer cachexia*, the dramatic weight loss caused by greedy cancer cells robbing the body of its nutrition. Cancer cells are more likely to thrive and grow when they have access to carbohydrates. They are far less efficient at utilizing fats as a source of nutrition. It makes perfect sense that slowly transitioning the dog's diet to one that is higher in fats and lower in carbohydrates will help fuel the *patient* rather than the cancer cells.

Dietary therapy is *not* effective as a sole treatment for your dog's cancer. Rather, its use is recommended in conjunction with other cancer-fighting therapies. Ask your vet to recommend an appropriate diet and source of omega-3 fatty acids (found in fish oil), which have been shown to inhibit the growth and spread of cancer. As with any diet change, remember that a slow transition is ideal lest you end up with an entirely different problem on your hands (one that involves professional carpet cleaners).

Slowly transitioning the patient's diet to one that is higher in fats and lower in carbohydrates will help fuel the patient rather than the cancer cells.

Making Decisions about Treating Your Dog's Cancer

First and foremost, I want to emphasize that whether or not to treat, and how best to treat your dog's cancer is truly a *personal* decision. With this disease, more than any other, we all bring a different set of values, ideas, and life circumstances to the table. Whereas one person might balk at the notion of only another six months of quality time, another might be relieved and overjoyed at the prospect. Whereas you might have the ability to leave a paying job in order to take care of your best friend during his recovery from cancer surgery, another might not have such flexibility. And, of course, sometimes the cost of care must influence your decisions. When it comes to treating cancer, there are no "right" or "wrong" decisions, only decisions that are best suited to you and your dog.

Cancer, whether human or canine, is emotional business capable of derailing clear-headed decision-making. In most cases, taking a few days to settle your emotions and do your research won't impact your dog's outcome in the least. By all means, ask your veterinarian how urgently your decisions need to be made.

When it comes to treating cancer, there are no "right" or "wrong" answers, only decisions best suited to you and your dog.

Your veterinarian will be able to tell you which treatment options are likely to be most effective against your dog's cancer. That's the easy part. What's not nearly so easy is determining whether or not such a treatment plan is reasonable for you and your dog. Here are some questions to consider. The answers will guide your decision-making:

❶ Is your dog's personality well suited to the recommended therapy?

If your dog is a pushover for anyone and everyone who gives him a pat on the head (and a cookie), he may relish the opportunity for weekly chemotherapy visits. On the other hand, if he turns into a quivering quaking emotional wreck as soon as you turn into the vet clinic parking lot, perhaps he'd be better suited to a chemotherapy protocol that involves coming in less frequently. You may not really know how your dog will respond emotionally until you've made your first few visits. Most dogs tend to become less anxious as time progresses, perhaps because of all the treats and attention they receive.

❷ Is your schedule flexible enough to accommodate the treatment recommendations?

Many chemotherapy protocols involve once-weekly visits for the first couple of months. Rarely are evening or weekend chemotherapy appointments available. If radiation therapy has been recommended, you may need to drive a consider-

QUICK REFERENCE

Questions to Ask Your Vet about Treating Your Dog's Cancer

- What is the name of the cancer?
- Where is the cancer located (anatomically speaking)?
- What experience do you have treating this kind of cancer?
- Do you anticipate the cancer can be cured?
- If the cancer cannot be cured, what is the likelihood of complete or partial remission?
- Is there an urgent need for me to make a decision? How much time do I have?
- What are all the treatment options?
- What are the potential risks and benefits associated with the proposed treatments?
- What are the logistics for each treatment under consideration? Will they necessitate outpatient or inpatient visits? How frequently will my dog require treatment and over what time period?
- What are the side effects of therapy and what can be done to minimize them?
- What is the average survival time with each treatment option?
- What is the likelihood of maintaining a good quality of life with the different treatments?
- Which option do you recommend, and why?
- What are the costs for the various treatment options?
- How do you determine whether or not remission has been achieved?
- What is the backup plan if treatment fails?
- If I choose not to treat, what do you anticipate happening?
- Should I do anything differently for my dog at home (activity level, diet, supplements, exposure to other dogs, vaccinations, heartworm preventative medication)?

able distance to a treatment facility. Oh, and did I mention that radiation therapy is typically administered Monday through Friday for *three consecutive weeks*. In some cases, you can board your dog at the radiation facility during the week and bring him home on weekends—but of course, being "away from home" might be a major source of stress for your dog (and for you).

❸ Is it financially feasible to proceed with the vet's treatment recommendations?

Be sure you are sitting down when you are given a cost estimate. Combination chemotherapy protocols, radiation therapy, and many surgical procedures are "big-ticket" items. Hopefully you have a "nest egg" tucked away, or medical insurance (see chapter 10, p. 230). When you have limited financial resources, I strongly encourage you to be very clear with your veterinarian about exactly what you can afford. Some chemotherapy is almost always better than no chemotherapy. So, don't nix the notion of chemotherapy altogether, simply because the "ideal protocol" isn't affordable. If your budget is limited, your veterinarian will be able to create a protocol that provides "the most bang for the buck."

❹ Is it emotionally reasonable for you to treat your dog's cancer?

For someone who has just experienced the ravages of cancer therapy, either for herself or a loved one, it may simply be impossible to consider cancer therapy for her dog (no matter what kinds of reassurances are provided). And that's perfectly okay. Remember, whether or not to treat your dog's cancer is a highly personal decision.

Who Should Treat Your Dog's Cancer?

Oncologists and Internists

When it comes to treating cancer, a specialist's input is recommended.

Veterinary oncologists devote their entire professional lives to the diagnosis and treatment of cancer. They attend continuing educational seminars so they can be up on the latest advances in testing and treatment. So who better to treat your dog for cancer than a board certified veterinary oncologist? Well here's the rub. At the time of publication, there are just over 200 such oncologists in the US, and roughly four million new cases of canine cancer diagnosed each year. You do the math! If these oncologists could clone themselves thousands of times over, they still wouldn't make it home in time for dinner.

With such a short supply of oncologists, it's natural that board certified veterinary internists are next in line to see cancer referrals. And internists, myself included, do indeed see a large cancer caseload (approximately 75 percent of my patients have cancer).

Both veterinary oncologists and internists can be found at most university teaching hospitals as well as in private specialty hospitals. Refer to the American College of Veterinary Internal Medicine (www.ACVIM.org) to find a medical oncologist or internist in your neck of the woods. The American College of Veterinary Radiology (www.ACVR.org) will help you locate a radiation oncologist.

Surgeons

When surgery is needed, you'll be hard-pressed to find an oncologist or internist eager to do the job. Most of us are so far removed from the operating room that we just might be the last veterinarians on earth you'd want to see holding a scalpel. Cancer surgery is tricky business, which is why oncologists and internists tend to refer their surgical patients to board certified veterinary surgeons. You can find one in your area by visiting the American College of Veterinary Surgeons Web site (www.ACVS.org).

Alternative Specialists

It is rare to find an oncologist or internist who has acquired significant knowledge in the area of alternative or complementary medicine. Clients interested in Chinese herb therapy, acupuncture, chiropractic, massage, homeopathy, or any other such treatment for their dog are generally referred to an expert with specialty certification in these areas. In fact, sometimes more than one practitioner gets involved because someone who performs chiropractic or homeopathy may not be well versed in Chinese herbs and acupuncture. You can find specialists in complementary and alternative medicine in your area by visiting www.IVAS.org (acupuncture), www.animalchiropractic.org (chiropractic), www.tcvm.com (acupuncture and Chinese herb therapy), and www.theavh.org or www.drpitcairn.com (homeopathy).

Many family veterinarians feel comfortable diagnosing, staging, and treating some types of cancer. Others prefer to refer to a specialist right from the get-go. Talk with your vet about *her* preferences, but remember *yours* are the ones that matter most. If you are feeling uncomfortable asking her for a referral, review the information in chapter 9, p. 207. Remember that meeting with a medical oncologist, radiation oncologist, internist, surgeon, or practitioner of complementary medicine in no way obligates you to choose one sort of diagnostic or treatment plan. You are committing only to gathering the information you need to make the best choice possible for your dog.

Talk with your vet about her preferences, but remember yours are the ones that matter most.

Final Words of Wisdom

By now, I'll bet I've worked with a few thousand people as they've struggled with whether or not to treat their dog's cancer. They've taught me a great deal that can help make this significant decision a bit easier for others. I'm passing along a few of the more important things I've learned with the hope that they will help get you on a path that's right for you and your dog.

- **Remember, you can call it quits at any time!**

When you agree to start, remind yourself that you are not signing a contract stipulating that you will continue treatment, no matter what. When you say, "Yes," to chemotherapy or radiation therapy, all you are committing to is the very next treatment. If your dog is becoming an emotional wreck, getting worse rather than better, or experiencing negative side effects, simply speak up and say that it's time to consider a different strategy. Here's the bottom line: if ever you don't like what you see, you can call it a day. Sometimes, simply knowing that this "out clause" exists gives people the wherewithal to give cancer therapy a try.

- **Keep things in perspective.**

Just because we can't cure a particular type of cancer doesn't mean it's not treatable. We treat lots of dogs with incurable, life-threatening diseases for the plain and simple reason that therapy often restores an excellent quality of life.

Take for example a dog with congestive heart failure. He'll never be cured of his disease and achieve his normal life expectancy, but medication has great potential to restore the quality of his life for a month to years. Treating cancer isn't all that different. Yes, for most people "cancer" sounds scarier than "heart failure," but each has the potential for a successful outcome with treatment. When it comes to survival times, don't forget to think in terms of "dog years" rather than human years. When your vet tells you that treating the cancer might buy another year of good quality time, consider that a year represents 10 percent of a 10-year-old dog's life! With that perspective, an extra year might sound like a tremendous proposition.

- **Focus on the quality rather than the quantity of life.**

When I counsel people about the potential pros and cons of treatment, I describe the three goals of cancer therapy as, "Quality of life, quality of life, and quality of life." There's no doubt in my mind that we have failed our patient enormously when we achieve "quantity" (longer life) only.

- **Enjoy the "honeymoon."**

Speaking of quality of life, even if your dog is doing fabulously well, it may be tough for you to enjoy his company when your mind is preoccupied with sadness and worry about the fact that your dog has cancer. Your happy, playing, wagging pup doesn't vanish just because cancer cells live in his body. He doesn't know he has cancer; he gets off easy when it comes to enjoying life and staying "in the moment." Let him serve as your psychological role model. Get the emotional support you need to take pleasure in the extra quality time you have with your dog. By the way, if he is acting glum, he may be responding to the way his favorite human is feeling!

- **Hope for the best, prepare for the worst.**

Veterinarians who have treated a good deal of cancer become relatively adept at predicting how their patients will fare. But, every one of us readily admits that we encounter some dogs that surprise the heck out of us. We've all had patients who crashed when we least expected it. Likewise, we've all witnessed recoveries that feel more like minor miracles than responses to our treatment. It just goes to show you that not every dog reads the cancer textbook! Truth be told, rarely can veterinarians look their clients in the eye and tell them that things are utterly hopeless. We simply don't know for sure how any one of our patients is going to respond.

So what's the take-home message here? I'm encouraging you to incorporate "hope" in all of your thoughts, feelings, and decisions about your dog's cancer. At the same time, avoid letting hope prevent you from recognizing when it is the end of the line. Your denial at this critical time might deprive your dog comfort and compassion at his time of greatest need.

- **As always, do your best to satisfy your peace of mind.**

I know I've discussed your peace of mind several times before in this book, but when it comes to cancer decisions, it bears repeating. There's no doubt that you and your veterinarian want what's best for your dog. The problem is, when it comes to treating cancer, it may be impossible to know what that is. When faced with this situation, I recommend that you base your decision, at least in part, on what best contributes to your peace of mind. I'm not suggesting that you ignore your dog and think only of your own needs. Rather, I feel this is just one more approach for tackling your decision-making dilemma.

Consider the following example: if your dog has been diagnosed with terminal cancer, the unfortunate certainty is that, barring any other catastrophic events or illnesses, his disease will ultimately be what ends his life. If treatment is successful, the length of remission time is anyone's guess.

Should you treat his cancer aggressively, conservatively, or not at all? If you are unsure, I encourage you to try this mental exercise.

Think about how the worst-case scenario for each of your options would play out in terms of your peace of mind. If you forego therapy and are at the brink of euthanasia in a month, will you be kicking yourself, wishing that you had pursued treatment? Likewise, if treatment was tried and failed a month later, would you have regret? When you consider all of these worst-case scenarios, which one would allow you the greatest peace of mind? Which will keep you awake at night? When I work through this exercise with my clients, I sometimes see an expression of relief wash over their face as they finally experience some clarity about how to proceed.

- **Follow your own heart.**

By encouraging you to follow your own heart, I'm recommending you stay true to your own values and feelings about dealing with your dog's cancer. Do your best to avoid being influenced by those who express shock when they learn that you are contemplating treating your dog's cancer. They may tell you that the appropriate treatment for cancer is "putting the poor dog out of his misery." And, when they find out how much you might be paying to have your dog treated ... well, it's probably best to avoid telling them in the first place how much you might have to spend!

Likewise, avoid those people who think that all dogs must be treated as aggressively as possible for anything and everything. I encourage you to wear a thick skin around such "influential" people (maybe take a sabbatical from socializing with them).

Provide yourself with plenty of open-minded friends who are interested in supporting you, rather than influencing you, as you wrestle with your decision. Remember, you know better than anyone else what is right for you and your best buddy.

VINNIE While writing this book, my beloved nine-year-old Golden Retriever, Vinnie, succumbed to cancer. I include this story as a way of letting you know that when it comes to making decisions about cancer, even veterinarians struggle to figure out what is best.

In February 2006, my husband (also a veterinarian) felt a firm lump on the top of Vinnie's head. A CT scan and biopsy of the lump revealed that Vinnie had a rather rare bone cancer called multilobular osteosarcoma. It was apparent as the mass continued to grow it would create pressure directly on the brain, eventually causing pain and profound neurological symptoms. Staging tests determined the cancer was confined to Vinnie's skull—not a trace of it anywhere else in his big golden body. The only treatment for a multilobular osteosarcoma is surgery. In Vinnie's case it was likely that most, but not all, of the tumor could be removed. Even if surgery was successful, at some point in time (how much time was anyone's guess), the mass would reappear.

We had quite the decision on our hands. Dare we subject our overtly healthy and exuberant dog to a surgical procedure that wouldn't be curative and might, in fact, do more harm than good? I had utmost confidence in the neurosurgeons I work with, but the fact that the surface of Vinnie's brain would be exposed and possibly traumatized during the surgery meant there was the chance of significant neurological complications.

Here's how we went about making our decision. we compared the certain and awful outcome without surgery to the possibility of prolonging a good quality of life with an operation. As scary as it felt at the time, there was little doubt that proceeding surgically best suited our peace of mind. I gave my neurosurgical colleagues the green light.

Fortunately, all went well, and although it was necessary to remove a very small portion of his brain along with the cancer, Vinnie had a complete recovery (some people might argue that a good deal of a Golden Retriever's brain can be removed without causing any noticeable change!) Vinnie did so well that a couple of months later he tore a cruciate ligament (in the knee joint) while performing his "Tigger routine."

So, Round Two of difficult decision-making: should we surgically repair the torn ligament, or simply restrict Vinnie's activity level knowing that the knee would remain a permanent source of discomfort? This was made all the more difficult by the knowledge that his skull cancer might reappear at any moment. We struggled whether or not it was reasonable to ask our dear Vinnie to endure another big operation. After significant soul-searching and a "conversation" or two with Vinnie, we again decided to proceed with surgery.

One of my gifted colleagues performed Vinnie's knee operation, and wouldn't you know it, again within a couple of months, he was back to his old self. The next nine months were, thankfully, medically uneventful. Vinnie enjoyed the ideal Retriever lifestyle—eating, playing, eating, sleeping, eating, swimming, eating, and receiving love and attention from his humans. Oh, and did I mention eating?

On September 4, 2007, we found our boy in an acute state of collapse, unable to get up on his own. He had appeared completely normal that morning. Our discovery was grim—Vinnie was hemorrhaging into the pericardial sac that surrounds the heart. The pressure of the blood within this sac was preventing his heart from contracting normally (imagine an ace bandage wrapped tightly around the heart muscle).

We knew that our luck with Vinnie had run out when an ultrasound exam allowed us to see the source of the bleeding. A tumor was growing from the right atrium of his heart—a different type of cancer than the one that had been removed from his skull. There was no medical or surgical way out of this. We nonsurgically drained the blood from the pericardial space with the hopes that we were buying ourselves some time for closure. We were able to bring Vinnie home, and when his symptoms recurred later that evening, we put our dear boy to sleep while he was eating his cookies, surrounded by his human, canine, and feline family.

"I went to the psychiatrist, and he says, 'You're crazy.' I tell him I want a second opinion. He says, 'Okay, you're ugly too!'"

—RODNEY DANGERFIELD

9

A Second Opinion Is Always Okay

In human medicine, a second opinion results in a new diagnosis as often as 30 percent of the time. There are no such statistics available in veterinary medicine, but as an internist who provides lots of second opinions, I have every reason to believe that the percentage is comparable.

A second opinion results in a new diagnosis as often as 30 percent of the time.

Second opinions serve two valuable purposes. The clearest benefit is for your pup. Although the health outcome may not be a positive one, another point of view typically facilitates a more expedient diagnosis and treatment. The other beneficiary is you—second opinions tend to be reassuring, and allow you to feel you are doing the best job possible for the dog you love so dearly. As you may have guessed, when it comes to significant health issues (regardless of species), I'm a strong proponent of getting more than one opinion.

The Courage to Talk to Another Doc

Intellectually, I think we all know that second opinions are a good thing. From an emotional point of view, there's a whole bunch of rea-

SWEET PEA From the moment I met Mary with Sweet Pea, her 12-year-old little terrier mix, I could sense her discomfort. Sweet Pea had been diagnosed with hepatitis two months prior, and Mary wasn't convinced that her precious dog was getting any better. But, Mary felt like she was "sneaking out" on her family veterinarian to come see me for a second opinion. She loved her vet and didn't want him to know for fear of hurting his feelings and possibly undermining their relationship.

As we began the examination, Mary told me that Sweet Pea was on a number of medications (she wasn't sure what they were), that three blood panels had been run to compare liver enzyme values (she did not have a copy of these results), and that X-rays (which she didn't have with her) had been taken the week before (Mary had been told they were abnormal, but she couldn't remember exactly what about them was wrong).

Are you getting a sense of the problems that arise from the "sneaky second opinion"? Without Sweet Pea's medical records, I would have had to start all over, something that wasn't in Sweet Pea's best interest, or that of Mary's finances.

After some cajoling, Mary acquiesced to let me contact her family veterinarian who, by the way, was most gracious. Once I had Sweet Pea's medical history, I was able to reassure Mary that the diagnosis and the recommended therapy were correct, and that Sweet Pea was indeed improving. Mary was so relieved! I could just see it written all over her face when she got the news she had hoped for.

A few months later, I met Sweet Pea's vet at a local meeting and he mentioned that our mutual patient was doing great, and thanked me for putting Mary's mind at ease.

sons why people cower at the mere thought of getting another vet's opinion. But, remember, you signed on to become a crackerjack medical advocate, and like it or not, getting second opinions is holding up your end of the bargain! Here are some of the common emotional barriers people must break through, and some coaching to help you come to your senses.

- **You don't want to hurt your vet's feelings. She has been there for you through thick and thin for so many years, and when your dog was hit by a car two years ago, she saved his life. How could you possibly doubt her?**

Your veterinarian is a big girl. If her feelings should be a little bit hurt, though it is unlikely they will be, she will get over it. Unless she's fresh out of vet school, this won't be the first time a client asked another vet for an opinion, and it won't be the last. Remember, your vet's primary concern should be your dog's health—not her own feelings. This is part of the professional oath of office we take when we graduate from veterinary school.

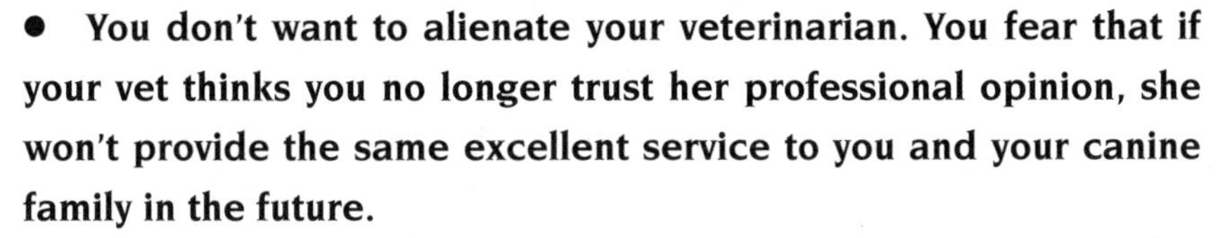

● **You don't want to alienate your veterinarian. You fear that if your vet thinks you no longer trust her professional opinion, she won't provide the same excellent service to you and your canine family in the future.**

If your vet responds negatively, then she has an ego the size of a hippopotamus! It is far more likely that she is certain you will continue to have her care for your menagerie because she has confidence in her ability and her relationship with you.

● **You're afraid the second opinion could be contradictory.**

Based on the new information provided by the second vet, it may be perfectly clear which assessment is correct. If, however, you still lack clarity, it behooves you to get a *third* opinion. Is getting a third opinion a pain in the #*@!? You bet it is! Is a third opinion in your dog's best interest? Most definitely!

● **You are concerned that the second vet is going to repeat all the tests your doctor already performed.**

The only reason to worry about this is when you are "sneaking out"—not telling your family veterinarian your intentions and consequently not providing Doc Number Two a complete medical record (chart notes, laboratory data, current medications, X-rays).

- **You fear it will take too much time and cost too much money.**

The only thing you are committing to is an office visit, and it may end up revealing a less time-consuming and less expensive option for diagnosis and treatment.

- **You think you will be obligated to continue working with the second veterinarian.**

A second opinion is simply that. You are, in no way, compelled to do anything beyond the consultation. Once you and your dog have obtained a second opinion, you can go back to your family veterinarian—or you can stick with Doctor Number Two. The choice is yours. Which vet do you feel more comfortable working with? Which one is better prepared and experienced in the diagnostics or treatment your dog needs?

- **Your dog has a condition requiring urgent medical care—you fear there isn't time to get more than one opinion.**

If the nature of your dog's emergency dictates that things need to happen lickety-split, it may not be possible to get a second opinion, unless there is another vet on the premises readily available to provide one. If you've done your homework in advance, you know which emergency facility gets your family vet's endorsement (see p. 50). Try to call her for advice. Indeed, she might want to be involved in the decision-making.

Sometimes it is possible to find an intern, resident, or clinician at a veterinary teaching hospital—local or even out of state—willing to provide you with a telephone consultation (use those time zones to your advantage).

• **You've been holding your veterinarian up on a pedestal, and you don't really want to admit to yourself that she may not be "perfect."**
Get over it! There is no such thing as an infallible veterinarian.

• **You're not really sure you want a bad diagnosis confirmed.**
Denial is not part of your job as an advocate. Get the help *you* need, so that you can get the help *your dog* needs.

WHEN to Get a Second Opinion

Now that you've got the gumption, how do you know it is time for a second opinion? Your veterinarian might recommend one, but often it will need to be your good idea. Listed below are some prompts meant to tap (or pound) you on the shoulder so you will begin to tell yourself, "Maybe I *should* get a second opinion."

• **You have a gut feeling.**
Gut feeling, sixth sense, your inner voice—whatever you call your experiential wisdom, pay close attention to it! Rarely do our gut feelings lead us astray, but ignoring them certainly can.

• **You simply want to be "more" certain.**
This is a perfectly fine reason. It means that you are being a shipshape medical advocate.

• **Your veterinarian has recommended a complicated surgery or other significant procedure.**
Just as in human medicine, when it comes to surgery or involved procedures, vets have varying levels of expertise. Never believe the phrase,

SECRET FOR SUCCESS

Remember, Your Veterinarian Isn't Perfect

I remember a particular client and patient with a good deal of embarrassment. A lovely gentleman asked me to evaluate his adorable Tibetan Terrier named Pirate because of a head-shaking problem. As I routinely do, I performed a thorough physical examination. In the process, I was surprised to discover an enlarged lymph node, possible evidence of a cancerous process. After we discussed this finding, I collected a small needle sample from the lymph node and sent it off to the lab. Fortunately, the enlarged node was found to be completely benign. When I called with the good news, the client expressed tremendous relief. He also ever-so-graciously informed me that he had taken Pirate to an emergency clinic the night before because his head-shaking had intensified, and a foxtail (a grass-type foreign body) was removed from deep within his ear canal. I was mortified. I had been so sidetracked by the enlarged lymph node that I'd failed to examine Pirate's ear canals. I completely neglected to address the problem that brought Pirate in to see me in the first place!

"First do no harm." This is the mantra that guides my professional life, and I'll bet that your vet feels the same. As much as we try to do no harm, the truth is that all veterinarians, myself included, have made—and will continue to make—professional blunders.

I am profoundly grateful to the savvy client who catches my error before any harm can come from it. I'm certain that the majority of my colleagues feel as I do. As your dog's medical advocate, you should remind yourself that your veterinarian isn't perfect (neither is the technician, receptionist, pharmacist, groomer, or anyone else who may be caring for your dog's medical needs). So when you recognize what you think may be an error or oversight, for your dog's sake, speak up as loudly as necessary.

"routine surgery" or "routine procedure." There is no such thing! Even the simplest surgery or procedure has the potential to cause significant complications, especially when general anesthesia is involved. Think about it, if you were in need of shoulder surgery, would you go to the doctor who does 100 shoulder surgeries a year or the one who does 10?

- **Your dog isn't getting any better or is getting worse.**

Some people give it a month, some a year, while others only a few

Ways to Talk to Your Vet about a Possible Error

"I'm so glad you discovered Casey's dental disease. No wonder he has such bad breath! Do you have a sense of why he's been limping?"

"The prescription vial says to give this antibiotic twice a day, but I think I remember you telling me that you were going to have me treat Lily three times a day. I just want to be sure I'm doing the right thing."

"I'm pretty sure that Stanley had vomiting and diarrhea when we gave him this medication a few years ago."

"I notice that these X-rays are of Cleo's left front leg. I think that it's her right front leg that is giving her trouble."

"Before you take those stitches out, I recall you telling me that we were going to leave them in for three weeks rather than two."

days before moving on and acquiring a second opinion. Do what feels right to you, and let your inner voice guide you.

- **Your vet doesn't specialize in the disease your dog has or the treatment your dog needs**.

If your own family physician discovered you had cancer, there is no question you would be referred to an oncologist—someone who works with the disease day in, and day out, and regularly pursues continuing education in her field. So, if the specialists your dog needs are not close by, it may well be worth the drive (or flight!) to find one (refer to the section on veterinary specialists, p. 215).

- **You want more information about alternative treatment options.**

Perhaps your vet is well versed in Western but not Eastern medicine, and you wish to know more about Chinese herb therapy or acupuncture. Ask your veterinarian for a referral to the right specialist, so that you can learn about the potential risks and benefits of *all* your options.

- **You've lost faith in your veterinarian.**

This happens from time to time—and it's no big deal. Do your research to find another veterinarian who will foster more confidence.

WHERE to Get a Second Opinion

The next question to tackle is where to go. Most people have plenty of options. I'm listing some of them on the following pages, along with some important pros and cons.

Professionals

❶ Another veterinarian within the same practice as your family veterinarian

Pros: This is most convenient option and may not cost much—if anything at all. It's sometimes possible for the second opinion to occur at the same time as the first. Say your dog has an unusual looking skin problem. Your family vet might be able to invite one of her colleagues to have a look and render a second opinion. How easy is that! And of course, by staying within the same practice, there is no need to transfer medical records.

Cons: Sometimes doctors who have been working together for a long time begin to think alike and practice medicine in a similar, if not identical manner, and this so-called second opinion is really just a glorified duplication of the first.

❷ A general practitioner in a different practice

Pros: This is a relatively convenient option, because you can usually choose someone close by, and the second doctor may have a collegial relationship with yours—it's always a good thing to have veterinarians collaborating about your dog's health. Such a process is often a source of good ideas.

Cons: You can end up with a doctor who has no more, and possibly less, expertise. And, your family vet and Doc Number Two may even have an adversarial relationship, thereby nixing the notion of any collaboration.

❸ An officially recognized veterinary specialist

This is a veterinarian who has completed three to four years of rigorous training following completion of four years of veterinary school, and who has obtained certification or an advanced degree in a particular area of expertise (see p. 79). As I've mentioned before, chances are your vet has established good working relationships with a variety of specialists in your area, and will be happy to refer you and call on your behalf.

You can also check specialty organization Web sites where lists of board certified specialists and their addresses are provided. If you are fortunate enough to have a vet school in your area, you are certain to find veterinary specialists hard at work there. When researching a specialist for your dog, make sure that the appropriate words or initials appear after her name (see sidebar, p. 218).

People often talk themselves out of going to see a specialist, thinking it will obligate them to proceed with complicated, involved, and expensive diagnostic tests or treatments. Fear not! A consultation with a specialist is nothing more than that—you have committed to the office visit *only*. You may choose to stay with the specialist for your dog's care, or the specialist's opinion may simply confirm the course of action your family veterinarian has recommended.

Pros: They are experts in their field. They very likely attend, and possibly even teach, continuing education in their area of specialization. They will arrive at a diagnosis more expediently, and be aware of state-of-the-art treatment options.

Cons: Significant travel time may be required. Specialists' fees are usually higher than those you'd find in a general practice.

Self-Help

Reference Books

Pros: It's convenient and relatively inexpensive to access dog health references in bookstores or public libraries (see some suggestions, p. 376).

Cons: For many topics, the books may have outdated or incomplete information.

The Internet

Pros: Easy, readily available, and inexpensive, for the judicious surfer, the Web is an efficient way to confirm the information you are getting from your veterinarian is on par with state-of-the-art medicine.

Cons: It is easy to be lead astray on the Web. There is a plethora of inaccurate data that is ripe with faulty assumptions and conclusions, plus anecdotes of miraculous cures, red herrings, and hocus-pocus. Veterinarians don't like to spend time dissuading clients from ridiculous notions obtained online. On the other hand, they love conversing with an educated, well-informed client. So, be discerning about the Web sites you visit: veterinary college sites reliably contain accurate information. Remember to Google the words "veterinary college" along with the subject you are researching. Your vet should also be able to refer you to specific sites that are accurate and helpful.

Alternative Sources

Pet Psychics/Healers/Communicators

Pros: The services provided by individuals advertising such skills are unlikely to be harmful and they can benefit client morale.

Cons: This may be a waste of your money that could be spent on a recognized specialist, but more importantly, it has the potential to derail or delay a more beneficial health-care strategy.

Second Opinion Etiquette

When paying a visit to a new veterinarian for a second opinion, it pays to heed some guidelines. To maintain the harmony between you and the rest of your dog's health-care team, I suggest the following:

1. As already mentioned, let your primary care veterinarian know that you are interested in a second opinion (see sidebar, p. 220). Express gratitude for the

QUICK REFERENCE

Veterinary Specialty Organizations

Organization	Specialty	Credential
American College of Veterinary Small Animal Internal Medicine (www.acvim.org)	Internal medicine	Diplomate, ACVIM
American College of Veterinary Internal Medicine, Cardiology (www.acvim.org)	Cardiology	Diplomate, ACVIM, cardiology
American College of Veterinary Internal Medicine, Neurology (www.acvim.org)	Neurology	Diplomate, ACVIM, neurology
American College of Veterinary Internal Medicine, Oncology (www.acvim.org)	Oncology (cancer medicine)	Diplomate, ACVIM, oncology
American College of Veterinary Surgeons (www.acvs.org)	Surgery	Diplomate, ACVS
American College of Veterinary Dermatology (www.acvd.org)	Dermatology	Diplomate, ACVD
American College of Veterinary Radiology (www.acvr.org)	Radiology	Diplomate, ACVR
American College of Veterinary Ophthalmology (www.acvo.org)	Ophthalmology	Diplomate, ACVO
American College of Veterinary Emergency and Critical Care (www.acvecc.org)	Emergency and critical care	Diplomate, AVECC
American College of Veterinary Anesthesiologists (www.acva org)	Anesthesiology	Diplomate, ACVA
American College of Veterinary Behaviorists (www.dacvb.org)	Behavior	Diplomate, DACVB
American College of Veterinary Nutrition (www.acvn.org)	Nutrition	Diplomate, ACVN
American Veterinary Dental College (www.avdc.org)	Dentistry	Diplomate, AVDC
American College of Theriogenologists (www.theriogenology.org)	Theriogenology (reproductive medicine)	Diplomate, ACT
American Board of Veterinary Practitioners (www.abvp.com)	General practice	Diplomate, ABVP
International Veterinary Acupuncture Society (www.ivas.org)	Veterinary acupuncture	CVA
Chi Institute (www.tcvm.com) (acupuncture and Chinese herb therapy)	Chinese veterinary medicine	TCVM
Academy of Veterinary Homeopathy (www.theavh.org)	Homeopathy	AVH
Professional Course in Veterinary Homeopathy (www.drpitcairn.com)	Homeopathy	AVH
Animal Chiropractic Certification Commission (www.animalchiropractic.org)	Chiropractic	AVCA
University of Tennessee Certificate Program in Canine Physical Rehabilitation (www.canineequinerehab.com)		CCRP (CCRT CCRA)
Canine Rehabilitation Institute (www.caninerehabinstitute.com)	Canine rehabilitation therapy	CCRP (CCRT CCRA)

QUICK REFERENCE

"Must Brings" to a Second-Opinion Appointment

❶ A copy of the complete handwritten or computer-generated medical record for your dog covering the past several years (not necessarily just pertaining to the problem at hand)

❷ All laboratory data

❸ All X-rays—either the image film itself or the images on a CD-ROM

❹ Your dog's vaccination history

❺ A legible copy of any ECGs (electrocardiograms)

❻ Clear images from any ultrasound studies

❼ A case-summary letter prepared by your family veterinarian

❽ All current medications

work she's already done—it pays to be nice! You want her to willingly share pertinent information about your pup's condition with the doctor supplying the second opinion.

❷ Arrive a little bit early for your appointment with Doc Number Two as there will be paperwork to fill out. This can be a 10- to 15-minute process, even longer if the receptionist is busy.

❸ Bring along a legible copy of all of your dog's recent and relevant medical records, which should definitely include all laboratory data, imaging studies—including X-rays, ultrasound evaluations, computed tomography (CT) scans, and magnetic resonance imaging (MRI) scans—ECG (electrocardiogram) tracings, and doctor's notes. It really helps if all of this material has been arranged in chronological order. Icing on the cake is a legible summary pre-

QUICK REFERENCE

Second Opinion Conversation-Starters

Here are some ways to broach the subject of a second opinion with your family vet:

"You've had so many good suggestions for us to try. Thanks for all of your efforts. It seems like Izzy's skin problem isn't completely going away, and we'd like to consult a dermatologist. Is there someone you can refer us to?"

"You've been so great helping Tula with her hip problems. We really appreciate it. Our friends think that acupuncture made a big difference for their dog, and we'd like to give it a try. Is there someone you can refer us to?"

"You've been great. Thanks for giving us so much helpful information. The surgery you're recommending for Brinker sounds pretty complicated, and this is a really big decision for us. We'd like to get a second opinion just to be certain how to proceed."

"We love working with you, and Sophie loves you. You're the best. Thanks for your help thus far. We've always been big believers in second opinions. Sophie's kidney problems seem so complex that we'd like to consult a specialist. Is there someone you might recommend?"

pared by your family vet. Remember, a stack of invoices is *not* a substitute for your dog's medical record.

❹ Bring along *all* of your dog's current and recent medications so the veterinarian can read the actual prescription labels. Just like human doctors, vets often have lousy handwriting, so details from a printed label are often more reliable.

❺ It is tempting to immediately tell Doctor Number Two everything Doctor Number One said. Instead, try to give the second veterinarian a chance to draw objective conclusions by asking her own questions.

SECRET FOR SUCCESS

Be a Successful "Squeaky Wheel"

Few people really enjoy being a "squeaky wheel." It requires extra effort, not to mention courage, and there is always the risk of being perceived as obnoxious. Let's face it—when it comes to dealing with health-care issues (or cell phone companies or governmental agencies) there are times when we all must make a bit of a fuss in order to be heard. In your role as your dog's medical advocate, you may well find yourself in situations where your dog stands to lose if you don't start "squeaking"! Consider the following:

- It's Wednesday. Your vet said she'd call on Monday—two days ago—with some lab results. You left a message yesterday, your dog is still sick, and you're scheduled to leave on a business trip tomorrow, but still no call back.

- You woke up this morning to sounds of retching and vomiting, and now your poor pup has refused his breakfast even with the addition of your scrambled eggs. He's never missed a meal in the eight years you've known him. You've called your vet but the receptionist insists that the next available opening isn't for two days.

- Your dog had a cancerous growth removed a year ago, and a new lump has recently appeared at the same site. Your vet examines the lump, and tells you she's quite certain it is benign. She recommends simply waiting and watching. Your "gut" tells you otherwise, and you know that you won't be able to sleep at night unless you know with absolute certainty that this new mass is truly harmless.

You owe it to your dog as well as to yourself to voice your concerns and be persistent in such situations. When I train office staff, I always explain that there are two types of emergencies: "real" or "client-perceived." Both are important and require immediate attention. In the first, we take care of the patient's physical well-being. In the second, we take care of the client's emotional well-being.

As a seasoned veterinarian, I've come to recognize two types of "squeaky wheel clients." Simply put, there are those who are successful and those who are not. Both are trying to be good medical advocates for their dog, but here's the difference: the successful individual is careful to maintain a harmonious relationship with the veterinary staff. Her mantra is "persistence, patience, and politeness." Her requests begin as softly as a whisper, and if necessary gradually escalate to a stern tone of voice. The unsuccessful squeaky wheel manages to alienate everyone in her path. She relies heavily on persistence while neglecting patience and politeness. Escalation in tone may

SECRET FOR SUCCESS CONT.

be impossible because there is stomping and shouting right from the start. Both may get what they need in the short run, but by protecting her relationship with the veterinary staff, the former safeguards her dog's future health care.

With this in mind, here are some pointers:

❶ Choose a veterinarian with whom you feel comfortable conversing and interacting. Your dog needs you to stand confidently at his side—you won't be able to be an effective advocate if you are intimidated by your vet.

❷ Always accompany your requests with sincere expressions of gratitude.

❸ Do your best to convey the emotional passion that is driving you to squeak! Let it be known your stomach hurts or you didn't sleep a wink last night from worry. Make your feelings apparent.

❹ Assess whether or not your request is truly reasonable. Sure, you'd like to pick up that prescription refill before noon, but it's possible the clinic staff is dealing with two or three emergencies on top of their already fully booked day.

❺ You may be fuming because you left three messages for your vet before you finally received a call back. When you do speak with her and address your concerns about your dog, it's perfectly fine to question why you needed to leave so many messages—but do so politely!

❻ Don't criticize your family vet. The new veterinarian might get the notion that you are a natural born complainer and you will likely be complaining about her next week—to a third individual. I can guarantee that, just as teachers talk with each other about their students, vets talk about their clients, especially the good ones and the bad ones!

QUICK REFERENCE

Successful and Unsuccessful "Squeaky Wheels"

Successful: "Hello, this is Mr. Patient. I know that you all must be extremely busy, and I appreciate your help. I'm really hoping to talk to Dr. Wonderful today about Tig's lab results. I'm really worried about him. Might there be a time that I could call back and speak with Dr. Wonderful today?"

Unsuccessful: "Hello, this is Mr. Impatient. I've left three messages for Dr. Unresponsive. Why haven't I heard back from him about my dog's lab results?"

Successful: "Hi, Dr. Wonderful. Thanks so much for taking the time to speak with me. I'm feeling worried because Jasper doesn't seem to be any better, and I'm wondering if we might try a different approach. What do you think we should do?"

Unsuccessful: "Hi Dr. Incompetent. I'm frustrated because I've already spent $600 at your office, and my dog is still sick! What are you going to do about it?"

Successful: "I know that you are really busy. Thanks for taking the time to speak with me. I've called because I'm feeling terribly worried about Elmo. I think he's declining quickly. Is there any chance Dr. Wonderful could see him today? I'd be able to come in any time that works for her."

Unsuccessful: "Listen, I don't care if you are busy—this is an emergency! I've been bringing my dog into your clinic for five years now, and I expect Dr. Overbooked to see him today!"

"My dog is worried about the economy because Alpo is up to $3.00 a can. That's almost $21.00 in dog money."

—JOE WEINSTEIN

10

Money Matters

When it comes to veterinary care for your dog, money is a fact of life, and I know of no one, myself included, who likes talking about it. It's a sensitive topic, one that is capable of stirring up distress, embarrassment, and anxiety.

Like most veterinarians, I typically feel some degree of awkwardness when addressing the "M" word. Talking "fleas" is one thing—having a candid conversation involving a client's bank account is another ball game! I invested vast amounts of time in learning medicine, but never received an ounce of education pertaining to client communication. I learned how to do the "money talk" by the seat of my pants, and I've certainly made my share of mistakes figuring out which approaches work and which ones don't. I came to recognize the benefit of having candid conversation and supplying cost estimates. I also try to avoid getting overly involved when helping people figure out how to pay their bill.

Talking "fleas" is one thing—having a candid conversation involving a client's bank account is another ball game!

I've learned that clients deserve to be made aware of all their options, whatever the size their piggy bank. When it comes to decisions involving money and four-legged family members, you just never know what people will choose to do. A wealthy person might be philosophically opposed to spending a lot of money on her dog's health needs, but by the same token, someone who is barely getting by might take on a third job in order to help her best friend get well.

Be completely up front and honest about your money concerns. This can open doors to other options making better financial sense.

Money Talk

There are typically two situations that prompt "money talk." The first revolves around the estimate, or how much things such as diagnostic tests and treatment are going to cost. The second occurs when it's time to pay the bill. Here are a couple of pointers for holding up your end of the conversation when discussing finances.

Wait for an Attentive Staff Member

If there is a need for "money talk," do your best to capture someone's undivided attention. All too often, a receptionist will want to go over an estimate or discuss a payment plan in between telephone rings and in the midst of a busy waiting room filled with people who can overhear every word. Let the staff member know if privacy is desired and that you are willing to wait for this and her full attention.

Honesty Is the Best Policy

You believe in and want the recommended tests or treatments, but your budget says, "No way." For everyone's sake, I encourage you to be completely up front and honest about your money concerns. This can open doors to other options that make better financial sense. In addition to discussing a plan for payment, perhaps a less expensive test can be run, or a more economical antibiotic can be tried.

People often have a hard time laying their financial cards on the table because they are fearful of being judged. What they may not realize is that well trained staff clearly understand that what a person can afford to pay for veterinary services is, in no way, a reflection of how much she loves and cares for her dog. Should you sense a lack of respect for your situation, I encourage you to ask to speak with your vet or another staff member, such as the front office supervisor or hospital administrator.

Price Shopping

70% of Americans surveyed consider a dog-friendly office an important job benefit.

I truly wish there was no need to discuss this, but for some dog lovers, price shopping for veterinary care is a fact of life. This may be driven by financial necessity, or just by the desire to get a "good deal." Are there veterinary clinics that offer excellent care and don't charge much for it? Yes, but you *must* do your homework to figure out which ones they are (see below). If you don't, I worry, for your dog's sake, that you might be sacrificing good quality care in exchange for a service that is less expensive.

Rather than bury my head in the sand, and ignore the topic of price shopping altogether (I'd make a ridiculous looking ostrich), I offer the following advice:

❶ **Price shop *only* when it is a financial necessity.**

❷ **Avoid sacrificing medical care quality.**

That old cliché, "You get what you pay for," is true. Be extremely thorough in your investigation; don't settle for brief, over-the-telephone quotes. Visit other veterinary clinics, tour the facilities, and meet staff. Hold all potential contenders to the same standards you used when choosing a family vet in chapter 3 (see p. 39).

❸ Watch for "hidden" fees.

Some clinics may offer an extremely reasonable quote for a procedure, but then charge an *additional* fee for the initial office visit or for post-surgical necessities like removing stitches.

❹ Keep in mind the potential for complications.

If a significant complication occurs due to below-par care, you will likely end up spending a great deal more money (not to mention associated emotional energy) treating it than you would have spent at the better, more expensive clinic to begin with.

❺ Talk with your family veterinarian.

Yes, I know that this sounds counterintuitive and you might be worried about insulting her. I say, nonsense! How could your vet possibly be upset if you tactfully explain that you have significant financial constraints and need to find the least expensive, quality provider of the recommended service? (The way you *can* cause upset is by "sneaking out" to a low-cost clinic, and then coming to your veterinarian with your tail between your legs when a complication arises.)

Your veterinarian wants what is best for your dog. She is the person most qualified to suggest more affordable options that will least compromise his quality of care. Given a list of other vets in your community, she will likely be able to tell you which ones offer high quality care and service. Because of your honest approach and good intentions, she may even be able to bend her own practice's payment rules.

Almost a third of dog owners say they would take a 5% pay cut if that meant their dog could accompany them to the office.

Cost Estimates

Many veterinarians automatically offer forth an estimate of costs, for any services provided beyond the basic office visit (see p. 59). If yours isn't in the habit of doing so, know that requesting an estimate is a perfectly appropriate thing to do—it is your right. How else can you possibly know if your bill will be $200 or $2,000? It benefits your veterinarian, too—she definitely wants you to be able to pay your bill! Note: *asking for a cost estimate in no way reflects how much you love your dog*. If you feel awkward requesting one, try these conversation starters:

"I'd like to get a cost estimate for your recommendations. I just want to be sure that I have the correct amount in my bank account."

"It would be great if I could get a cost estimate. I never like surprises!"

"Your recommendations make a lot of sense. A cost estimate would be very helpful for me to know if we can get started."

Get It In Writing!

It's always best to get an estimate in writing, especially when the service involves more than one test or treatment. There are so many ways one can misinterpret a verbal estimate. Clients who want to save me time and trouble often suggest, "Oh, just give me an estimate off the

There are so many ways one can misinterpret a verbal estimate.

top of your head." I invariably decline, because the top of my head is capable of producing some rather wild guesstimates! I insist on writing down details because I learned long ago that, when I do give off-the-cuff figures, I almost always lowball the price. I think it's because in my heart of hearts, I truly do want it to be affordable.

Understanding the Estimate

If you don't understand various items, ask for an explanation. If you are considering two different treatment options, request written estimates for both. Also, confirm just how far into treatment the estimate takes you—suppose the estimate covers two days of hospitalization, but your vet told you it could be as many as four. The difference in expense between the two is likely worth knowing up front.

Remember, it is unreasonable to expect the final bill will be an exact replica of the estimate.

Sometimes an increase in cost occurs because of a change in circumstances; for example, a blood transfusion might be unexpectedly needed during surgery. It's the staff's responsibility to let you know about any such "change orders," ideally before the extra charge is incurred. Remember though, it is unreasonable to expect the final bill will be an exact replica of the estimate. If it's a bit lower, hurray! A bit higher? This, too, should be expected.

Financing Your Dog's Veterinary Care

Veterinary clinics' and hospitals' overhead costs are constantly increasing, and the advanced medical technologies now readily available result in increased fees for services provided. For example, the MRI scan provides a wealth of information compared to its less expensive cousin—the plain old X-ray. And, hip replacement surgery is a fabulous, though pricey, option when your four-legged friend is lame with arthritis and anti-inflammatory medications aren't getting the job done.

Consequently, many clients are surprised by the amount of money needed to care for their sick dog because they may not have considered costs beyond those needed for their pup's routine care. Plus, human health insurance has trained us to think about "co-pays" rather than the "total pays" that most veterinary hospitals require.

So, how will you manage to pay for your dog's unanticipated illness or injury? Advance planning is the ticket. Whether you open up a medical line of credit, buy pet insurance, or stuff money under your mattress, it pays to be prepared. If you are superstitious like me, you can take comfort in knowing when you plan for the worst, the worst is far less likely to happen!

Know Your Payment Options

When it comes to ways to pay your bill, there is no standard veterinary clinic menu, though most accept cash, credit/debit cards, and checks. Do your homework to find out which ones apply at your facility. When you research payment options *before* vet services are provided, trust me, the staff is most appreciative. What receptionist hasn't heard, "What do you mean, you don't take Discover®?" or, "I assumed you offer a multiple-pet discount."

When you research payment options before vet services are provided, trust me, the staff is most appreciative.

Many, if not most, clinics and hospitals require payment *in full* at the time services are rendered. Even when the dog is insured, the insurance company reimburses the client, not the veterinarian, so the client must pay up front (see p. 232). Some clinics offer payment plans, but they are not obligated to do so. A few vets may be willing to barter or even accept "hold checks"—it never hurts to ask. It also helps to bear in mind that a deposit might be requested before diagnostics or treatments are initiated.

A Line of Credit for Medical Expenses

A payment option called CareCredit® is offered by a growing number of veterinary facilities throughout the US and Canada. This program has been successfully employed in the human medical and dental fields since 1984, helping people pay for any out-of-pocket medical expenses, including elective procedures such as dental care, cosmetic procedures, and corrective eye surgery.

In a nutshell, CareCredit® provides a line of credit that can be used to pay for veterinary expenses. When payment is due, CareCredit® pays the veterinary hospital, and the cardholder repays CareCredit®. There is no annual fee, and no obligation to use the product. For example, you can obtain CareCredit® when you first get your puppy, but not use it for years until significant medical expenses arise.

There are various plans to choose from that allow for interest-free payments for a pre-determined period of time (however, high interest fees often do apply if a balance is not paid off by a certain date). Clinic staff can get you enrolled, or you can sign up online or by telephone (the process is as easy as obtaining a department store credit card). Find out if your clinic offers this unique payment option, and visit www.carecredit.com for further information.

Two-thirds of dog owners say they would put in longer hours if they could bring their dog to work.

The World of Pet Health Insurance

Veterinary health insurance has been around for a good long time, but only recently is it achieving greater popularity with consumers. My sense is that its growth initially was stymied by inadequate, "slow-pay, no-pay" reimbursement policies. This seems to be changing, and some providers are now willing to give greater reimbursement amounts to policy holders, thus attracting people who want to take advantage of high-end diagnostic and therapeutic options that might otherwise be unaffordable.

Deciding whether or not to purchase a medical insurance policy for your dog requires serious consideration. And, if you decide to go ahead, figuring out which insurance company is the best fit can be daunting. I'll help you sort through these decisions on the following pages.

Quality veterinary care is expensive, and as the cost of living increases, so, too, will the cost of doing business with your vet.

Although it is considered to be far less necessary than human medical insurance, should your dog suffer some sort of catastrophe—such as being hit by a car—pet insurance might be your best, if not your only, way of financing his recovery. Without question, quality veterinary care is expensive, and as the cost of living increases, so, too, will the cost of doing business with your vet. At the time of publication, a surgical repair of a torn cruciate ligament—a common knee injury in large breeds—costs $2,000 to $4,000. The average fee for an MRI scan (including general anesthesia) is $2,000 to $3,000. Treating diabetes can cost several thousand dollars over the span of a dog's lifetime.

How Pet Insurance Pays

Remember, when it comes to pet insurance, third party payments are the exception rather than the rule. This means that the veterinarian receives payment directly from you, the client, and not from the insurance com-

SECRET FOR SUCCESS

Naming your dog "Lucky" is just asking for trouble.

Enough said!

pany. *You are still responsible for paying your veterinary bills*. The insurance company then reimburses you as per the terms of your policy.

To Insure or Not To Insure

As you ponder whether or not health insurance makes sense for you and your dog, consider the following questions:

- **What are your financial resources?**

A new puppy means multiple examinations, vaccinations, deworming, heartworm preventative, and spay or neuter surgery—expenses that will need to become part of the household budget. If an emergency—illness or accident—occurred, could you pay for your dog's recovery? Think about the types of expenses you might encounter, such as surgery, an ultrasound evaluation, hospitalization with or without intensive care, and specialist consultations. Could you absorb such costs should the need arise tomorrow? How do these numbers compare to the amount needed to purchase a year's worth of medical insurance for your pup?

Your six-month-old Golden Retriever may be the picture of health, but how about several years down the road when he's become a "golden oldie"? Perhaps purchasing and maintaining pet insurance when your dog is young makes sense. This way, you can rest assured there will be no exclusions for pre-existing conditions (see p. 235), and you may have the option of "locking in" a lower premium rate.

- **Are you inclined to take the "do-everything-possible" approach when it comes to treatment?**

The price tag for aggressive veterinary care is considerably higher than for conservative approaches. If you answered this question affirmatively, insurance might be well worth the investment.

- **What best suits your peace of mind?**

Will you sleep better at night knowing that, no matter what happens, insurance will allow you to pay for excellent, top-of-the-line care? Or, will you lie awake fearing that you are just throwing money away with yet another insurance policy that might never be needed?

Time to Do Your Homework

At the time of publication there are at least a dozen pet insurance companies competing for your dollars. Be prepared to dedicate some serious time to comparing providers. Start by soliciting advice from your veterinary clinic staff. They are quite likely to hear from clients who are happy or unhappy with their insurance company dealings! Then investigate the recommended companies and pay attention to the way your top contenders deal with the following issues:

❶ Pre-existing conditions

A pre-existing condition refers to any health problem your dog developed prior to the date you purchased his health insurance policy. Some insurance companies permanently "blacklist" such pre-existing conditions (it will never be covered by their insurance policy). Others will provide coverage for the pre-existing problem once it has been resolved, or after a clearly defined disease-free time interval. Try to get a sense of how liberally or conservatively the pre-existing condition policy is interpreted. For example, say your pup had a seizure after eating snail bait when he was six months old. Now, at four years, he has developed epileptic seizures. Would coverage for the epilepsy be declined because of the prior seizure event, even though the seizure then and the seizures now are completely unrelated?

Be aware that when you file a claim, companies typically "subpoena" all prior medical records with the intent of identifying pre-existing conditions. How they interpret these records is anybody's guess. I'd love to be able to tell you that logic always prevails, but my experience tells me otherwise. Hassles over pre-existing conditions make a strong case for insuring your pup when he's fresh out of the womb, before he's had a chance to have anything go wrong.

❷ Repeat offenders

If you are wondering, "What are these?" consider the following: I live in northern California. If you think that life in this area is pretty darned perfect, you haven't heard about *foxtails*—awful little bristly weeds that grow rampantly throughout the late spring and summer months. They seem hellbent on finding their way into dogs' noses, ears, eyes, mouths, and just about every other orifice. Not only is the dog's body incapable of degrading or decomposing them, the foxtails are barbed in such a way that they can only move in a "forward" direction. Left untreated, they migrate through the body causing infection and tissue damage, and invariably, veterinary intervention is necessary to remove the things. Well, some dogs are just downright predisposed to getting into foxtails. Vets treat these "foxtail magnets" year after year, and sometimes multiple times during the course of a year—they are a good example of repeat offenders!

How does an insurance company handle the dog that manages to get into the same mess requiring vet attention over and over again? Some providers have a yearly "cap" (maximum amount paid out) for each disease or situation, and every year the dog starts with a clean slate. Other companies enforce a lifetime cap, meaning that reimbursement for a specific problem might be permanently cut off long before your dog becomes a senior citizen. Of course, the ideal policy is one that imposes *no limits*—each new episode is covered. Be aware of this when shopping for insurance, because someday your adorable, "innocent" pup may become a repeat offender!

❸ Inherited diseases

The sad fact is that most purebred dogs come with a laundry list of potential breed-associated diseases. For example, many Dalmatians have an inherited metabolic defect that causes the formation of bladder stones, and affected dogs require a rather pricey treatment regimen, consisting of surgery to remove the bladder stones along with lifelong preventive medication and monitoring. Most, if not all, insurance providers exclude this Dalmatian-specific problem from coverage. If a purebred dog is part of your family, learn the medical issues he is predisposed to, then compare them to the insurance company's list of breed-specific exclusions. Note: If you encounter reluctance to provide such a list, it's probably best to move on to a different company.

❹ Anticipated conditions

Compare insurers' coverage of medical needs that you know are likely to arise. If you are planning to breed your female dog, consider expenses associated with pre-breeding screening for inherited diseases, ovulation timing, pregnancy, and whelping (delivering her pups). Big dogs are more prone to hip dysplasia (arthritis) and gastric torsion (twisting of the stomach, requiring immediate surgical intervention). Smaller breeds are notorious for developing heart-valve issues and severe periodontal disease. See the sidebar on p. 20 for tips on how to predict your dog's possible medical needs in the future.

❺ Policy coverage options

Most insurance providers offer a variety of coverage beyond "the basics." You might want to choose a "wellness" policy that provides reimbursement for your puppy's office visits, vaccinations, and neutering. If your best friend is a senior citizen, look for an option that covers routine geriatric screening. Check to see if the company allows for policy changes during the course of your dog's life.

❻ Claims processing

If you tend to have a tight budget, ask the provider what the average lag time is between filing a claim (money you've already paid the vet) and receiving the reimbursement check. Determine how claim disputes are handled. Keep in mind your vet will *not* want to get involved in insurance company disputes. We all know "people doctors" who have jumped ship because of having to deal with insurance providers. Veterinarians are trying to stay off that road to professional burnout!

Question-and-Answer Time

Plan some serious online time to get answers to your questions, many of which will come from the insurance company's Web site. Invariably, some questions will need a "human's" attention, so create your own list or use those provided in the sidebar on p. 240, and contact customer service by phone, as well.

When phoning, note how long you wait to talk to someone—a living, breathing human being, that is. Was he knowledgeable and personable? Treat this interaction with the customer service representative as though you are conducting a job interview, which in fact, it is! The insurance company is interviewing for the critical job of insuring your dog's health.

Ask the customer service representative to walk you through a couple of hypothetical claims. Use the sample vignettes I've provided below, or make up your own scenarios. Why do I want you to do this? For years, unsatisfactory reimbursement practices and illogical assessments of pre-existing conditions have been major bugaboos in the pet insurance industry. You should know in advance how your pup's pre-existing medical conditions will be interpreted. You need an insurance company that uses a clear-cut, objective, predictable formula when calculating claim reimbursements. The answers should be readily forthcoming and provided confidently. If, instead, you encounter a good deal of hemming and hawing, and answers do not include specific dollar amounts, it's time to move on to the next company on your list.

And lastly, consider this surefire tactic (okay, perhaps it's a tad devious). Present the exact same hypothetical cases to more than one rep in the same company's customer service department. If the responses are identical, you can be confident the insurance provider has a plan and intends to stick to it.

Hypothetical Scenarios to Present to Potential Pet Insurers

Scenario One

Your perfect little Poodle became abruptly paralyzed in her hind legs because of intervertebral disk disease (a slipped disk). Your family doctor referred you and your dog to a board certified neurologist. Thankfully following an MRI scan, surgery and rehabilitation therapy, your best friend is now acting like a princess again.

QUICK REFERENCE

Questions to Ask Pet Insurance Companies

About the Company

❶ How long has your company been in operation?

❷ Is coverage provided where I live?

❸ Is coverage provided when I travel?

❹ Can we utilize any veterinarian and hospital?

❺ Do you offer multiple-dog discounts?

❻ What are your customer service hours?

About the Policies

❶ What are the coverage plan options? (Most providers have multiple options; often the more that is covered, the more expensive the insurance premium becomes.)

❷ What are the deductible options? (As with many insurance policies, there is usually a required "out-of-pocket" investment before the insurance company provides reimbursement. Often the lower the deductible, the more expensive the policy.)

❸ How much will my premium increase every year?

❹ Is it possible to lock in the premium rate so that it doesn't increase?

❺ Will there be a penalty if I change my plan or deductible?

❻ Is preapproval of medical services ever required?

❼ What are the policy limits? Is there an annual or lifetime cap for a particular medical problem?

❽ Will I be able to find out the amount of insurance reimbursement before authorizing my veterinarian to proceed with recommended tests and treatment?

❾ Is the reimbursement amount for a particular service always the same, or does it vary based on the veterinarian's fee?

❿ Can my dog be dropped from coverage? If so, what are the criteria for doing so?

About Claims

❶ What is the process for filing a claim?

❷ How are claims handled when there is no diagnosis, either because tests were not performed or the results were inconclusive?

❸ How long does it take to receive payment on a claim?

❹ How are claim disputes handled?

❺ What is the specific formula for calculating reimbursement on a claim? Is there a benefit schedule (standard amount paid for a service regardless of what the veterinarian charges), or is it based on a percentage of the cost?

❻ Is there a maximum amount paid per incident (medical event)?

❼ Is there a maximum amount of reimbursement per calendar year?

❽ Is there a maximum amount of reimbursement per disease?

What's Included and What's Not

❶ May I see the list of diseases excluded from coverage?

❷ May I see the list of breed-specific diseases excluded from coverage?

❸ How much time must pass before a preexisting condition is finally covered?

❹ Are complementary/alternative medicine services covered?

- Homeopathy?
- Acupuncture?
- Acupressure?
- Massage therapy?
- Chinese herb therapy?
- Chiropractic?

❺ Are services associated with breeding, pregnancy, Cesarean section, and newborn care covered?

❻ Are consultations with specialists covered?

❼ Are second opinions covered? Third opinions?

❽ Is after-hours emergency care covered?

❾ Are costs associated with treating behavioral issues covered?

❿ Are costs associated with well care/preventive care covered?

- Vaccinations (including serology)?
- Heartworm testing and preventative?
- Spaying and neutering?
- Dental work?
- Geriatric screening?
- Flea and tick control?
- Microchip identification implantation?
- Deworming?

Here's the breakdown for what you paid for diagnostics and treatment. How much of these out-of-pocket expenses will be reimbursed?

Visit with family veterinarian	$75
Visit with neurologist	$150
MRI scan	$2,500
Surgery	$4,000
Hospitalization	$1,000
Rehabilitation therapy	$500
Total	**$8,225**

Scenario Two

Your lovely Labrador ate the fish you just caught. Unfortunately, he swallowed the whole thing, hook, line, and sinker, leaving the fishhook embedded in the lining of his stomach (confirmed with an X-ray). Your family veterinarian anesthetized your boy, and attempted to remove the fishhook by way of endoscopy (a long telescope device that is passed down the esophagus into the stomach). No luck, darn it. The fishhook was too deeply embedded. So your Lab was wheeled into your vet's operating room where the fishhook was surgically removed. He was then transported to a 24-hour emergency/critical care facility for overnight monitoring. He went home the next day, tail wagging the whole way—ready to go on another fishing trip! How much of the following will be reimbursed?

Visit with family veterinarian	$75
X-rays	$125
Endoscopy procedure	$400
Surgery	$1,500
Overnight care	$600
Total	**$2,700**

Scenario Three

Your magnificent mutt developed allergic dermatitis (itchy skin). Your family veterinarian referred you to a board certified dermatologist who performed skin allergy testing. Medicated shampoo and allergy desensitization injections were prescribed. Fortunately, they created a huge improvement. As recommended, you've been taking your pup back to the dermatologist every two months for recheck visits. A search of his medical records revealed that five years ago he had a mild allergic reaction to one of his puppy vaccinations (before you had health insurance).

How much will you be reimbursed? (The key here is, will allergies be considered a pre-existing condition based on the vaccination reaction?)

Visit with family veterinarian	$75
Visit with dermatologist	$150
Skin testing	$600
Medicated shampoo (one year)	$150
Desensitization injections (one year)	$400
Recheck visits (one year)	$400
Total	**$1,775**

The Money-Savvy Advocate

Financial planning is a necessary part of being an advocate and is best done when your pup is in perfect health. When he becomes ill, money is the last thing you'll want to be thinking about. It doesn't matter whether you've purchased a health insurance policy, opened up a medical line of credit, or kept a bank account specifically for your dog's health needs—in the end you *will* sleep easier knowing that money matters won't get in the way of your dog getting the medical attention he needs.

"And beloved master, when I am very old and the Great Master sees fit to deprive me of my health, do not turn me away from you, rather take my trusting life gently, and I shall leave you, knowing with the last breath I draw, my fate was always safest in your hands."

—FROM "A PET'S PRAYER"
BY BETH NORMAN HARRIS

11

Euthanasia: Making the Best Out of a Difficult Situation

Unfortunately, sick dogs rarely simply pass away peacefully and comfortably in their sleep. In most cases their human family must face the issue of euthanasia. We are forced to become extraordinary medical advocates for our dogs as we grapple with one of the most difficult decisions of our life.

If you are in the midst of this gut-wrenching experience, I hope the information in this chapter will guide and support you. If you have already made the choice to put your dog to sleep, here you will learn how to achieve the best possible outcome—a gentle, peaceful, and comfortable process for your pup, and peace of mind for you and others whose life was touched by your wonderful dog.

Who Should Make the Decision?

No one is better equipped to make the decision about humane euthanasia for your dog than you.

No one is better equipped to make the decision about humane euthanasia for your dog than you. You know his habits and expressions and the things that bring him joy and comfort. It doesn't matter if you are young or old, or living with your

CORI was a sweet and gentle Siberian Husky, and she responded to my medical poking and prodding by giving me lots of kisses. Cori's "mom," Maureen was also a delight to work with, and she loved her dog dearly. They were constant companions. The first time we talked, she told me that Cori had been a birthday gift from her late husband. Maureen and her husband shared the first three years of Cori's life together, and after her husband passed away, Cori became Maureen's main reason for carrying on.

Unfortunately, when I ran tests on Cori's enlarged lymph nodes, I found a type of cancer called lymphosarcoma. Chemotherapy chased the cancer away for nine wonderful months, during which time Cori felt great. She and Maureen visited the beach daily (Cori's favorite thing to do).

But the cancer returned, this time resistant to chemotherapy, and Cori's quality of life began to deteriorate. Maureen and I talked daily as she struggled with the difficult prospect of euthanasia. She didn't want Cori to suffer, but she couldn't bear the thought of life without her very best friend and emotional link to her late husband. After much soul-searching, they made one last trip to the beach, and later that day, Maureen sat on the couch in my office with Cori draped across her lap as I administered the euthanasia solution.

Afterward, Maureen told me the only reason she could finally let go of Cori was because of the dream she'd had the night before. In the dream, she saw her late husband kneeling down and Cori running into his outstretched arms.

very first dog or your eighteenth. When it comes to deciding what is in your dog's best interest, *you* are in the driver's seat. Be wary of the well-meaning coworker or relative who tries to sway you by stating, "If he was my dog, I'd put him to sleep." Be cautious of the vet who offers euthanasia as the only option. Likewise, don't be influenced by your friends at the dog park who advise against euthanasia under any circumstances. Firmly remind yourself that he is not their dog! He is *your* pup, *your* buddy, *your* constant companion. You are the one with the most insight about the quality of his life.

Deciding whether or not to put a beloved dog to sleep is never without angst. The easy way out is to succumb to a state of emotional distress and allow someone else to decide. Don't do it! People who relinquish responsibility seem to have the greatest difficulty moving through the grieving process. They experience lingering doubt and inordinate guilt, wondering if the decision was made prematurely, or if euthanasia was the right choice at all. So try to stay tough, and continue to struggle with the task at hand—your dog needs you to do this. It's payback time for all of that companionship and unconditional love your buddy has provided you over the years.

When Is the Right Time?

"How will I know when it's time?" is one of the most common questions I'm asked when clients are contemplating euthanasia. They may have just received a hopeless prognosis or witnessed a significant decline in the quality of their dog's life. By asking me this question, they clearly want to avoid acting too early. What they sometimes forget to think about, however, is that the greater error may be in waiting too long. In some situations, putting it off can result in needless suffering for their dog and long-lasting emotional turmoil for the family. There is no "best" or "correct" time. The challenge is figuring out when the time is just right for you and your dog.

When people contemplate putting their dog to sleep, I'm often asked if I think he is in pain. I can tell they are hoping my answer will clearly point their way. Unfortunately, it's not that simple. With the availability of many excellent pain-control drugs for dogs, pain may not be the most life-limiting issue. Conversely, if the dog is not feeling pain, it cannot be assumed he is not suffering. A dog with advanced kidney failure is not hurting, but his lethargy, vomiting, and decreased appetite undoubtedly suggest he is suffering.

Ask your vet to help you determine if the quality of your dog's life is diminished, whether because of pain or another type of suffering.

When I'm working with someone who is struggling to know when it's the right time, I ask the following questions:

- **Does your dog still respond enthusiastically to the things that would normally excite him, such as the jingle of car keys, the sight of his well-worn tennis ball, dinner time, the rumble of the UPS truck in the driveway, or the mention of his favorite word (the one you have to spell out when you don't want him to get excited)?**

- **Do the good days still seem to outnumber the bad?**

- **When you get down on the ground and go eyeball-to-eyeball with your dear companion, do you still see that familiar spark in his eye that lets you know that he wants to keep on going?**

- **Do you sense your dog is "hanging in there" and putting on his game face in order to take care of you? Your always-loyal best friend may feel that he doesn't have "permission to pass away" because you, his most beloved human, aren't quite ready to let go.**

Now, mix the answers to these questions with your gut feelings. Recruit support and guidance from trusted, open-minded friends and your vet. Ponder, process, question, and wrestle with all you are thinking, and repeatedly check in with that wonderful dog of yours. Trust your dog to tell you what you need to know, and trust yourself to know what is best for him. Believe it or not, at some point, the decision will become clear.

Trust your dog to tell you what you need to know, and trust yourself to know what is best for him.

SECRET FOR SUCCESS

Always Have a Contingency Plan

Several professional "nightmares" are dreaded by veterinarians everywhere. Here is one of them. A dog has just been hit by a car and is brought to the hospital by the pet sitter. The dog's back is broken, the prognosis is terrible, and if surgery is to have any chance of working, it needs to be performed right away. The poor pup is in tremendous pain. The only two reasonable options are surgery and euthanasia. Unfortunately, the person who would normally make such a decision is out of the country and cannot be reached. The pet sitter has received no authority to make such a monumental decision.

It's hard to imagine that something like this could ever happen to you and your dog, but for your dog's sake and your veterinarian's, please plan for this possibility. If you are going to be unreachable for a period of time, assign to someone the responsibility of making crucial medical decisions for your dog. This can be your vet herself, a trusted friend or relative, or the pet sitter caring for your pup while you are away. Get her assurance that she will be willing and available to take on such a commitment, then put it in writing (see a sample form on my Web site, www.speakingforspot.com), and leave a copy with the pet sitter, the person responsible for making decisions in your absence, and the veterinary office staff.

"Closure Time"

"Closure time" refers to the invaluable time spent between a dog and his favorite humans prior to euthanasia.

Within the framework of this chapter, "closure time" refers to the invaluable time spent between a dog and his favorite humans prior to euthanasia. Closure time provides the dog well-deserved pleasure and happiness at the end of his life. It creates precious memories for the family members, and feeds and nurtures their soul, giving them strength to withstand the final goodbye, and the subsequent grieving process.

When the end of your dog's life approaches, I encourage you to create some time for closure. Work with your veterinarian to find a way to make this happen. In the event of a sudden crisis or accident, this period of closure might be just a few private moments in the hospital setting. In the case of a chronic illness, it may last for days, weeks, or even months.

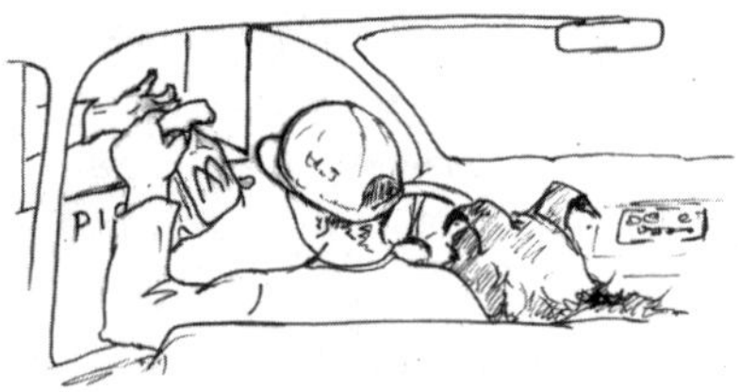

I always enjoy listening to my clients talk about how their closure time was spent. All kinds of decadent things happen, such as frequenting the McDonald's® drive-thru, ignoring the leash law at the beach, throwing away all those cans of diet dog food, and chasing the neighbor's cats. When my clients tell me such fun, crazy stories, they are usually sporting a grin, and I feel like laughing and crying at the same time. They've watched their dog experience fun and joy at the very end of his life.

Once you've made the decision to let your pup go, carve out this ever-so-important time for just the two of you. You'll be glad you did.

Choices to Consider

I cringe when I hear that a dog was "dropped off" at the vet clinic for euthanasia. I find myself wondering whether or not the now bereft person knew that she could have been by her dog's side. Many people don't realize that when it comes to the euthanasia process, they have many choices. Unless your dog must be put to sleep urgently, I encourage you to consider the following:

Who Will Be Present?

Most of my clients choose to be with their dog, something I always encourage for two reasons. First, I sometimes hear regret and guilt voiced by people who chose not to be there; those who remained with their dog are invariably glad they did. Second, unless the patient is unconscious and unaware, I believe in my heart of hearts that dogs derive profound security and comfort from the presence of their beloved humans. However, if the thought of being present at the vet's office feels emotionally overwhelming, know that it's okay; you are definitely not alone in the way you feel. Consider asking someone in your family, a friend, or a vet staff member to fill in on your behalf; someone who will provide lots of petting, loving, and cookies.

If you wish to be with your dog when he is euthanized, by all means do not take "No" for an answer!

If you *do* wish to be with your dog when he is euthanized, by all means do not take "No" for an answer! I view a veterinarian's refusal to honor this request as unacceptable, and recommend you find a more progressive vet. Remember, you only have one chance to get it right.

The question of whether or not children should be present arises. Certainly this is a matter of personal preference, but keep in mind, most kids under the age of 10 are capable of causing a major distraction. They might be alarmed or scared witnessing the euthanasia procedure, especially if Mommy or Daddy is upset. It is perfectly reasonable to discuss what is going to happen with children over the age of 10 or so, and allow them to determine if they wish to be present.

An option for children (as well as the client and other friends and family) is to be outside the room when the actual euthanasia is performed, then allowed to spend some time with their dog afterward. Whatever your decision, I cannot overemphasize the importance of being up-front with your kids about their dog's euthanasia. Children can be helped to understand the respect, love, and humaneness behind such a decision. Kids who don't hear the truth have a knack for knowing they are being lied to, and are sometimes left wondering if their dog is forlorn and lonely somewhere, especially when told their dog "ran away from home." There's lots of literature available to help you and your children with this process (see p. 376).

Is it beneficial for other family pets to be present? Might it help them understand why their constant companion and playmate is no longer at home? Honestly, I don't know if dogs or cats derive benefit by being there and "investigating" their buddy after he's passed. I do think that it is fine for other pets to be in attendance as long as the patient enjoys their company and they are not a source of unwanted distraction for the people involved.

After the euthanasia has been performed, I tell my clients that we are in no hurry to have them leave. I want them to feel comfortable and to spend as much time as they need to say goodbye. This is often a profoundly peaceful period when people feel a sense of relief at the gentleness of the procedure as well as the fact that their beloved dog is no longer suffering.

Where and How Should It Take Place?

A vet's office is the last place on earth some dogs want to be. If your dog falls into this category, be reminded that euthanasia doesn't have to be performed in the clinic setting: many veterinarians will make a house call to perform euthanasia and are also willing to transport the remains for cremation or burial in a pet cemetery if "home burial" is not desired or possible.

A vet's office is the last place on earth some dogs want to be.

Not every veterinarian, though, feels comfortable administering the euthanasia solution outside the controlled, vet clinic setting. In this case, I recommend contacting a "house-call" vet who is accustomed to performing the procedure in a variety of places. House-call doctors may be unavailable on an emergency basis, so when possible, advance planning is advised.

In addition to home and clinic settings, euthanasia can also be performed in the family car. It may sound a little bit crazy, but for some dogs the family car feels like a second home, and it is a reasonable choice for a dog that becomes a nervous wreck the minute he sets foot inside the vet office. It can be a tight fit, but if it makes the dog more comfortable, and the veterinarian is willing, I'm all for it.

If euthanasia is going to take place at the clinic, learn the specifics of where you and your dog will be. You want a room that is private and quiet where you can feel free to talk, cry, or meditate for as long as you like, both before and after. In my hospital, we have the luxury of grieving/visiting rooms that look nothing like examination rooms. They are carpeted, with sofas, rocking chairs, and dimmer lights. Feel free to add personal touches: Perhaps you wish to include religious or spiritual content. Maybe you want to spoon-feed Häagen Dazs® to your pup during his final moments. You might want to lie on the floor together, or if at home, sprawl in the grass in the backyard. Or, your dog might be happiest in his raggedy old dog bed, holding a soggy tennis ball in his mouth as he passes away. Make it as meaningful as you like, all the while focusing on your dear dog's comfort. Not every veterinarian will be keen on such unique ideas, but you never know until you ask.

If the need for euthanasia arises abruptly, you and your dog may wind up at an emergency care facility. Rest assured that the staff in these practices is very used to this situation. Nothing needs to change in terms of how you advocate for your dog during this most important time.

Note: I recommend handling all necessary paperwork (the bill, authorization for care of remains) beforehand. The last thing you want to do afterward is spend time taking care of business in a crowded waiting room.

How Will the Remains Be Cared For?

Some people have the means to accommodate a home burial for their dog. If this is unfeasible or (undesirable), other options include burial at a pet cemetery, communal cremation (more than one dog cremated at the same time), and individual cremation with or without return of the ashes. Most veterinarians have standing arrangements with pet cemeteries and crematoriums so your dog's remains can be picked up directly from the clinic, although I've had clients who wished to transport their pet's remains to the crematorium or cemetery to be certain that everything went as planned. Your vet will know the available options.

When individual cremation is your choice, find out how the ashes will be returned. Most commonly they are sent in a tightly sealed wooden box, which may not be the best choice if you intend to spread them.

Consider whether you want your dog's collar to stay with him or if you wish to retain it as a keepsake. Should you decide to keep it, leave it on until after your pup has been euthanized to avoid causing him unnecessary worry. Likewise, if you wish to gather a lock of silky hair or make a paw-print impression in clay, do these things afterward, rather than before.

What Actually Happens When a Dog Is Euthanized

Before I perform euthanasia, I always explain what is about to happen to my clients in order to dispel inaccurate preconceived notions. I want them to be reassured that it is a peaceful, pain-free process.

When a dog is euthanized, the vet administers an overdose of an anesthetic agent. Some veterinarians place an intravenous catheter before the euthanasia solution is given, while others prefer to inject the solution directly into a vein. Once the injection has been made (sometimes even before the injection has been completed), there is a quick, quiet loss of consciousness that resembles a dog being anesthetized for a medical procedure. Most people are surprised at how quickly this occurs—typically within 10 to 20 seconds. Soon thereafter, breathing stops, followed by cessation of audible heart sounds (your doctor will likely confirm this via a stethoscope). Because of muscle relaxation, emptying of the bladder or bowel may occur. The dog's eyes typically remain open, contrary to what most people anticipate.

Sometimes, the veterinarian and client prefer the dog to be sedated prior to euthanasia in order to alleviate any anxiety he may experience. This is perfectly acceptable, but keep in mind that the impact of sedation on a sick dog can be highly variable. Most people want to feel connected to their dog during his final moments, but there's a chance your dog may become unresponsive as a result of the sedation. Be sure to let your veterinarian know how you feel about sedation.

Advance knowledge of these predictable aspects of euthanasia usually prevents feelings of surprise or discomfort for those present. My sense is that, managed correctly, euthanasia is a rapid and pain-free experience for the dog. In fact, if he has been struggling to breathe or suffering from pain, the peacefulness that immediately follows is profound.

Alternatives to Euthanasia

Some people are opposed to humane euthanasia for moral, philosophical, emotional, or religious reasons. For years, such clients created quite a conundrum: my professional obligation to ease patient suffering did not seem compatible with my clients' convictions. Fortunately, with time, I've become a bit smarter and now have the means to reconcile what is best for the patient when the client is opposed to euthanasia. What a relief! The answer to this moral/ethical dilemma is hospice care—an alternative to euthanasia that emphasizes pain control for the terminally ill dog and emotional support for both the patient and the family. The practice of hospice care has certainly been embraced in human medicine, and it is beginning to catch on in the veterinary community.

Now, hand in hand with any discussion about euthanasia, I discuss this option, too. Hospice care is usually carried out in the dog's home. A great deal of time, energy, and focus is required of the patient's human family as they learn to provide the necessary supportive care and pain management. This may involve round-the-clock turning from side to side, carrying and bathing, assistance with urination and bowel movements, preparation of special diets, administration of medication for pain or nausea, and supplemental fluids, often by way of injection. Unlike their human counterparts, canine patients cannot clearly communicate the need for pain control. The caretaker must learn to recognize when pain is present. Elevations in heart rate, respiratory rate, or blood pressure; restlessness; and vocalizing (whimpering, whining, moaning) can all be indicators of pain. Your vet and her staff can teach you monitoring guidelines specific for your dog and his disease.

Providing hospice care is emotionally and physically draining; however, it can be a richly rewarding endeavor, especially when those involved believe that pain and suffering are being adequately managed. Terminally ill dogs are quite capable of "lingering" so it is not uncommon for hospice care to last for weeks, or even months. Some people decide to provide hospice care for a finite period of "closure time" and then euthanize; others opt for hospice care until death occurs naturally.

BOOMER I'm including the following story to remind you that veterinarians also struggle when it comes to end-of-life questions for their canine family members.

During my sophomore year in veterinary school, I fell in love with Boomer, an unbelievably adorable mix of Bassett Hound and Black and Tan Coonhound. As a puppy, his body was tiny in comparison to his ears, and, when on the run, his feet would get all tangled up with those ears—he ended up doing more somersaulting than running! He tolerated my long hours in veterinary school, and let me practice what I learned on him. He gave me the thumbs up when I met my soon-to-be husband, and Boomer moved with me from New York to California, wagging his tail the whole way. He spent his years frolicking, knocking over garbage cans (including the neighbor's), running after scents that only a true hound could detect and, of course, howling!

When Boomer was 11 years old, he developed an inoperable brain tumor, and I watched my sweet boy deteriorate until he was no longer able to move about. He continued to eat and wag for me, but only when coaxed. We'd been through so much together. I couldn't bear the thought of losing him. My two oldest children, Mickey and Jacob, were six and four years old at the time, and I was so hoping Boomer would be with them all through elementary school. Furthermore, Lily, our Golden Retriever puppy was only six months old, and she still had so much to learn from her "big brother."

The decision to put Boomer to sleep was one of the hardest I've ever had to make. I held him in my arms as my husband (also a veterinarian) administered the injection. It was a peaceful, quiet passing.

We opted to bury Boomer on our property. Before the burial, my husband, children, and I went in different directions, each of us looking for something special that would be meaningful to leave with our wonderful dog. I came back with Boomer's leash, Mickey arrived with his dog food bowl, and my husband brought along his favorite squeaky toy. We hollered out to Jacob, but there was no sign of him. After several minutes, Jacob finally arrived with his "good idea." We laughed and cried when we saw our four-year-old boy with our Golden Retriever puppy in tow.

"I have sometimes thought of the final cause of dogs having such short lives and I am quite satisfied it is in compassion to the human race; for if we suffer so much in losing a dog after an acquaintance of ten to twelve years, what would it be if they were to double?"

—SIR WALTER SCOTT

Appendix I

Common Symptoms and Questions Your Vet Will Ask about Them

In this section I have listed commonly observed symptoms that should prompt a visit to your veterinarian and the questions your vet is likely to ask about them. In addition, there are four key questions she will *always* want you to answer:

1. **How long has the symptom been present?**
2. **Has the symptom improved, worsened, or stayed the same since the time you or others first noticed it?**
3. **Has your dog ever had this symptom before?**
4. **Have you observed any other symptoms such as vomiting, weight loss, coughing, sneezing, or change in appetite, activity level, thirst or bowel movements?**

By contemplating the above questions—as well as those listed in conjunction with specific symptoms on the following pages—in advance, you and your family members will be able to compile important observations and information before you visit the veterinary clinic. Your vet will be most appreciative, and ultimately, your dog will benefit immensely from your efforts.

Symptom: Abdominal distention

Overview Abdominal distention refers to belly expansion or enlargement. It can be caused by: pregnancy; obesity; fluid or blood accumulation; organ enlargement; tumors; weakening of the abdominal muscles; and overdistention of the stomach or urinary bladder. If your large, deep-chested dog (built like a German Shepherd or Great Dane) has sudden abdominal distention, waste no time getting to your vet. Dogs with this body conformation are prone to *gastric torsion* (bloating with gas followed by twisting of the stomach), which requires emergency surgery. Even if your dog seems to be feeling great, abdominal distention is a definite reason to have your dog evaluated.

Questions Your Vet Will Ask You

If you have an unspayed female dog, is it possible that she was bred? When was her last heat cycle?

You may have thought your fence was tall enough, but when a dog is in heat, the impossible becomes possible!

Has your dog possibly eaten anything unusual?

A dog that just busted into the neighbor's garbage or a 40-pound bag of dog food is bound to have some abdominal distention (serves him right!) Take a minute to check around your house and yard (maybe the neighbor's, as well) before visiting your vet. The more she knows about what may have been ingested, the more likely she can help your dog.

Associated Symptoms to Report Change in thirst or appetite, decreased activity, vomiting, diarrhea, labored breathing, weight loss, straining to urinate or have a bowel movement

Symptom: Bad breath (halitosis)

Overview Dental disease is the most common cause of bad breath—also known as halitosis. Other disorders within the mouth, throat, back of the nose, windpipe, stomach, and esophagus (the muscular tube that leads from the mouth down into the stomach) are also capable of causing bad breath. Sometimes severe kidney disease, liver disease, and diabetes produce an unusual "metabolic" odor to the breath.

Questions Your Vet Will Ask You

Has your dog been eating normally?

Eating gingerly or favoring soft foods suggests a problem within your dog's mouth.

Has there been a recent change in diet?

A new diet may cause your dog's breath to smell different.

Has your dog had itchy skin?

Excessive self-grooming (especially near the anal region) has the potential to alter a dog's breath (and not for the better).

Has your dog had his teeth cleaned before?

If your dog has had dental plaque and tartar cleaned in the past, report if it improved your dog's breath at that time.

Associated Symptoms to Report Difficulty swallowing, burping/belching, vomiting, nasal discharge, coughing, sneezing, reverse sneezing (see p. 272), itchy skin, decreased appetite, lethargy, weight loss

Symptom: Behavior change

Overview Dogs are creatures of habit. They tend to wake up, eat, and expect their walks at the same time, day after day. So, when your dog doesn't "show up" at the usual place at the usual time or does something truly out of character, pay close attention. His altered patterns may be the result of an underlying medical problem. For example, a normally well house-trained dog that suddenly begins urinating in the house is more likely to have a bladder infection than a grudge about something.

QUICK REFERENCE

"Watch List" of Symptoms that Warrant a Visit to the Vet (Note: this list is not all-inclusive)

Abdominal distention	p. 259
Alopecia	p. 268
Bad breath	p. 260
Behavior change	p. 260
Blindness	p. 262
Blood in the stool	p. 266
Blood in the urine	p. 278
Collapse, inability to rise	p. 263
Cough	p. 264
Decreased appetite	p. 265
Decreased stamina	p. 266
Diarrhea	p. 266
Discharge from eyes	p. 274
Flatulence	p. 267
Hair loss	p. 268
Halitosis	p. 260
Inability to rise	p. 263
Incoordination	p. 268
Increased appetite	p. 269
Increased frequency of urination	p. 278
Increased panting	p. 269
Increased thirst	p. 270
Itchy skin	p. 271
Labored breathing	p. 272
Lethargy	p. 272
Limping (lameness)	p. 273
Loss of balance	p. 268
Lump/bump	p. 273
Nasal discharge	p. 274
Ocular discharge	p. 274
Pain	p. 275
Pruritus	p. 271
Regurgitation	p. 280
"Scooting"	p. 276
Seizure	p. 276
Sneezing	p. 278
Straining to have a bowel movement	p. 266
Straining to urinate	p. 278
Urination abnormalities	p. 278
Urine leakage	p. 278
Vaginal discharge	p. 279
Vomiting/regurgitation	p. 280
Weight loss	p. 282

Questions Your Vet Will Ask You

Does the abnormal behavior occur at a particular time of day?

Personality changes associated with severe liver disease are typically most apparent right *after* mealtime. Those caused by low blood sugar are at their worst right *before* mealtime.

Do your dog's hearing and vision seem normal?

Sudden deafness or blindness can produce dramatic behavioral changes—for example, a normally enthusiastic dog becomes subdued or fearful.

Do you get the sense that your dog is experiencing pain?

Dogs that are uncomfortable often become quiet and reclusive (think how *you* deal with a major headache).

Have there been any changes within the household?

Have you gone back to work after years of being home full time? Is there a new dog, cat, or baby in the house? Certainly these types of emotional disruptions are capable of causing behavioral changes.

Associated Symptoms to Report Incoordination, weakness, vomiting, diarrhea, change in thirst or appetite, aggression, lethargy/dullness, change in vocalization, restlessness/anxiety

❊ Don't Forget!
Check p. 259 for four key questions your vet will *always* want you to answer.

Symptom: Blindness

Overview Dogs are so adept at using all of their senses, rarely is partial blindness apparent to us. In fact, if the loss of vision has come about gradually and the dog has stayed in familiar surroundings, even complete blindness may go unnoticed. Causes of blindness include: glaucoma; cataracts; inflammation within the eyes; retinal disease; and neurological disorders that interrupt the normal nerve pathways between the eyes and the brain. A careful examination of the eyes often reveals the cause, but that may be only part of the story. For example, cataracts may be the reason for the blindness, but diabetes may be the cause of the cataracts.

Questions Your Vet Will Ask You

Did the blindness appear abruptly or gradually?

Think back over the past few months. Has your dog been more tentative than normal? Has he been working harder to find the cookie tossed his way? Has he seemed reluctant to go up and down the stairs, or go for walks, especially when it's dark?

What has prompted you to think that your dog has lost his vision?

For example, say you've seen his pupils turn from black to grayish/whitish in color. Dogs over the age of 10 develop a normal aging change in their eyes called *lenticular sclerosis*. Although lenticular sclerosis causes no real vision problems, people often assume they are seeing cataracts, and consequently, that their dog is losing his vision.

Associated Symptoms to Report Squinting, discharge from the eyes, change in size of the eyes, pupil size asymmetry, incoordination, weight loss, change in behavior, thirst, or appetite

Symptom: Blood in stool See Diarrhea, p. 266

Symptom: Blood in urine See Urination abnormalities, p. 278

Symptom: Collapse (inability to rise)

Overview Weakness, paralysis, syncopal episodes (fainting), and seizures are all capable of causing a state of collapse. It can sometimes be difficult to determine the underlying cause, especially if the symptoms are transient. Your observations will help your veterinarian hone in on the diagnosis.

Questions Your Vet Will Ask You

Did your dog lose consciousness?

It can sometimes be difficult to tell if a dog is having a seizure or a fainting episode. Seizures (see p. 276) typically cause the dog to become unconscious (complete lack of responsiveness), while a dog that has fainted will typically respond to loud noises or familiar human voices. Seizures and fainting have entirely different causes and treatments, so being able to report what actually happened is essential.

How long did the episode of collapse last?

If the collapse is caused by a heart rhythm abnormality or a seizure, the episode is usually short. Other diseases often produce a more prolonged state of collapse.

Did you see it happen or just find your dog in a collapsed state?

If you simply happened upon your dog already like this, you may have missed the primary event, such as a seizure.

When did the collapse occur in relationship to mealtime?

If your dog has low blood sugar or severe liver disease, food intake can directly impact the timing of the collapsing episode.

What was your dog doing at the time he collapsed?

Let your veterinarian know whether or not your dog was doing something exertional.

If your dog has had more than one episode, have they all resembled one another?

When the cause of the collapse is a seizure, in most cases, repeat episodes will be consistent in how the dog appeared before, during, and after the event.

Did the collapse appear to affect your dog's entire body or just his hind end?

A dog with *intervertebral disk disease* (a slipped disk) or some types of heart disease might experience a hind-end-only rather than a full-body collapse.

Was your dog's body stiff or flaccid during the state of collapse?

Seizures are often associated with rigidity. Other causes typically produce a more flaccid state.

Did your dog urinate, have a bowel movement, or salivate excessively while collapsed?

Any of these concurrent symptoms suggests that the cause of the collapse was a seizure.

Did your dog's gum color or heart rate change?

In the heat of the moment, you may not have made these observations (and if your dog is having a seizure, you really don't want to lift his lip to have a look at his gum color for fear of being bitten). But, if you happened to observe pale or bluish gums, or feel a pounding heartbeat, be sure to mention this.

Associated Symptoms to Report Incoordination, weakness, behavior changes, decreased stamina, labored breathing, abdominal distention, vomiting, diarrhea, decreased appetite, weight loss, increased thirst

Symptom: Cough

Overview A cough in a dog sounds much the same as it does in a human, and the causes are just as numerous. Allergies, infection, heartworm disease, tracheal (windpipe) collapse, heart disease, and cancer are some of the more common reasons a dog would cough.

Questions Your Vet Will Ask You

Has your dog had recent exposure to other dogs?

"Kennel cough" is an "umbrella term" used to describe contagious viral or bacterial *tracheobronchitis* in dogs. Let your vet know if your dog's cough began shortly after a visit to the groomer, dog park, boarding facility, or any other place where dogs congregate.

Does the coughing occur more at any particular time of day or night?

Coughing associated with heart failure is often at its worst after a dog has been lying down for a while. So, it makes sense that this cough is most persistent first thing in the morning.

Is your dog able to exercise normally?

If not, significant heart or lung disease may be the cause of the coughing.

Has your dog had exposure to smoke, dust, or any other inhaled irritants?

Time to 'fess up if you or someone else in the household is a smoker. Also let your vet know if the coughing began in conjunction with a move to a new house or exposure to new carpeting, bedding, or anything aerosolized (carpet cleaners, pesticides).

Does your dog's cough recur seasonally?

If your dog has been a chronic cougher, note whether it occurs more during a particular time of year. If so, the cough may be a result of seasonal allergies.

Does the cough sound wet or dry?

If the cough sounds wet (productive) observe any mucous or discharge spat up (assuming your pup doesn't just swallow it).

Has your dog consistently received heartworm preventative medication?

If you live in heartworm territory be sure to let your vet know if your pup's heartworm prevention has lapsed.

Are any other dogs in the household coughing?

If they are, this would suggest a contagious cause for the coughing, not to mention some sleep-deprived humans!

Does your dog have exposure to foxtails?

During the summer months in the western US, inhalation of a foxtail (a pesky plant awn) is always a consideration when a dog begins coughing abruptly.

Associated Symptoms to Report Labored breathing, lethargy, decreased stamina, discharge from the eyes or nose, decreased appetite, weight loss, loss of normal pink gum color

Symptom: Decreased appetite

Overview Dogs tend to be creatures of habit. Left to their own devices, they eat in a predictable fashion, day in and day out. Some dogs are gluttonous—others tend to eat based on hunger. If your dog's appetite diminishes or his tastes suddenly change (for example, after years of eating dry food, he'll now only eat canned food laced with table scraps), consider this a big red flag saying, "Hey, human, something is wrong with me!"

Although a decrease in appetite is a significant symptom, it is extremely nonspecific and can accompany hundreds of different diseases. Answers to the following questions will help whittle down the list of possibilities.

Questions Your Vet Will Ask You

Has there been any recent change in diet?

If you've just opened a new bag of dog food (even if it is the brand your dog is used to eating), it is possible that the formulation of the food is different or, worse yet, the food is somehow tainted. If you've just switched to a different brand or type of food, the taste difference could be the source of your dog's appetite change.

Have the household dynamics changed?

A dog that is hungry, but wimpy, may not eat if there is a new dog guarding the food bowl. Sometimes dogs go on a hunger strike when their favorite humans aren't around as much. Emotional turmoil affecting family members might curb the appetite of the sensitive dog.

Does your dog initially show interest in the food or is he completely disinterested?

This vital tidbit of information will be tremendously useful to your vet. Excitement about food that quickly vanishes suggests that he is experiencing pain or nausea associated with eating.

Does your dog's ability to swallow seem normal?

When your dog drinks water or takes a bite of food, does his swallowing motion appear exaggerated or uncomfortable?

Is your dog more inclined to eat softer foods?

This may be an indicator of pain associated with eating.

Have there been any recent changes in your dog's medications?

Many medications are capable of snuffing out a dog's appetite. If a drug or supplement dosage was changed or a new one started around the time you noted an alteration in appetite, be sure to mention it.

Associated Symptoms to Report Vomiting, diarrhea, lethargy, weakness, labored breathing, coughing, sneezing, increased thirst, weight loss, evidence of pain or discomfort

Symptom: Decreased stamina

Overview Does your dog "quit" after three tosses of the Frisbee whereas, before, he'd play all day? Does he seem to tire halfway through his walk? Decreased stamina or exercise intolerance is most commonly caused by heart, lung, or musculoskeletal disease. Be sure to see your vet should this symptom arise.

Questions Your Vet Will Ask You

When your dog appears fatigued, do you notice any other significant changes?

Be sure to mention and describe any observed labored breathing, pounding or rapid heartbeat, weakness, or abnormal gum color (blue, purple, or gray).

How long does it take for your dog to recover?

Does he go back to playing normally after a short rest or is he done for the day?

Has your dog been on heartworm preventative medication?

Exercise intolerance is one of the earliest symptoms of heartworm disease.

Does he appear unusually stiff or sore?

This would point to sore muscles or joints.

Associated Symptoms to Report Lethargy, decreased appetite, weight loss, vomiting, diarrhea, coughing

Symptom: Diarrhea

Overview Veterinarians use the term, "diarrhea" whenever there is increase in frequency or fluid content—even stools that are formed but softer than normal are considered diarrhea. It's easy to assume diarrhea equates with gastrointestinal disease, but this isn't necessarily the case. A number of other organ-system problems (kidney, liver, pancreas) can all cause diarrhea.

Questions Your Vet Will Ask You

What are the changes in your dog's bowel movement or bowel movement behavior?

Let your vet know when you observe any blood, mucous, change in color, parasites, or undigested food. Is your dog having excessive straining or gassiness when he has a bowel movement? Have his bowel movements become much more frequent, or are they occurring inside rather than outside? All these observations will help your veterinarian know if your dog's diarrhea originates from his small intestine or his large intestine. This makes a difference in terms of diagnostic testing and how your dog may be treated.

Has there been a recent change in diet?

Diarrhea can be a normal response to new food, especially if the transition was made abruptly.

Did your dog eat something he shouldn't have?

Check the garbage cans, the compost pile, your house and yard area (perhaps the neighbors' as well). Given the opportunity, most dogs have rather nondiscriminating tastes! Most any type of garbage is capable of causing diarrhea ("garbage-can gut"), but some food, such as raw chicken or fish, is capable of causing a more serious, infectious diarrhea.

Has your dog been losing weight?

If your dog is getting thinner in spite of a normal appetite, this suggests his diarrhea might be originating in his small intestine where nutrients are normally absorbed.

Has more than one dog in the household been having diarrhea?

If so, an underlying infectious or parasitic cause would be suspect.

Associated Symptoms to Report Change in appetite, lethargy, increased thirst, vomiting, change in appearance of the anus and surrounding area

⁂ Don't Forget!
Check p. 259 for four key questions your vet will *always* want you to answer.

Symptom: Flatulence

Overview Flatulence is a fancy way of referring to "passing gas." Some gas in the gastrointestinal tract is normal, but if your dog is "farting" up a storm, consider the following information.

Questions Your Vet Will Ask You

Has there been a recent change in diet?

A change in diet, especially one made abruptly, is capable of causing flatulence.

Are there any changes in the appearance of the stools or your dog's bowel movement behavior?

If you've observed diarrhea, straining, or pain associated with passing a bowel movement, or blood or mucous in the stool, your vet will know the primary problem is within the large intestine.

Have you observed burping or heard gurgly gastrointestinal noises?

Both of these concurrent symptoms suggest excessive gas within the gastrointestinal tract.

Associated Symptoms to Report Decreased appetite, vomiting, weight loss, lethargy, abdominal distention

Symptom: Hair loss (alopecia)

Overview The medical term for hair loss is *alopecia*, which can vary from complete baldness to thinning of the coat to poor regrowth of hair following grooming. Causes of alopecia include: skin infection (fungal, bacterial, parasitic); hormonal imbalances; and inherited skin disorders. Interestingly, chemotherapy rarely causes hair loss in dogs.

Questions Your Vet Will Ask You

Has your dog been itchy?

The hair loss may be a result of your dog scratching and biting himself. Itchiness, also known as *pruritus,* can suggest certain underlying diseases.

Are any humans or other pets at home itchy, too?

Sarcoptic mange and fleas are happy to jump ship to another species!

Does your dog's hair readily come out on its own?

If clumps come out with routine brushing, leaving bald spots behind, a hormonal imbalance could be suspect.

Associated Symptoms to Report Increased thirst, increased appetite, excessive panting, lethargy

Symptom: Incoordination (loss of balance)

Overview If your normally agile dog is now clumsy, his head is cocked to one side, or he appears dizzy, it's time to pay your vet a visit. Various toxins, as well as low blood sugar, and diseases of the brain, spinal cord, or inner ear are capable of causing incoordination.

Questions Your Vet Will Ask You

Has your dog had exposure to any toxins?

Snail bait, antifreeze, and certain mushrooms are just a few of the items with toxic substances capable of causing incoordination. Be sure to scout out your dog's home environment (inside and out) for anything he might have consumed.

Is your dog taking any medication?

Some are capable of causing incoordination, even when given at an appropriate dosage.

Has your dog had an ear infection?

Sometimes infections within the external (outer) ear canal are associated with inner ear problems, resulting in a loss of balance.

Associated Symptoms to Report Change in behavior or personality, lethargy, weight loss, change in thirst or appetite, pain

Symptom: Increased appetite

Overview This symptom is often ignored because most people figure that a hearty appetite means a healthy dog. This is not always the case. If your dog has recently become a major "chowhound," or he's developed an appetite for unusual things such as plants or dirt, a trip to the vet is recommended. Causes of increased hunger may include: certain types of gastrointestinal diseases; hormonal imbalances (diabetes mellitus, Cushing's syndrome); increased expenditure of energy; and behavioral issues.

Questions Your Vet Will Ask You

Has there been a change in his food?

Perhaps your dog is eating more because he likes the taste better, or you are feeding a weight-reduction diet, and he is hungrier because he's getting fewer calories.

Is there new competition for food?

Sometimes the nibbler will become a gobbler when a new canine face shows up! It is often the case that in a two-dog household, one dog eats his dinner more quickly than the other, then moves in for a "second helping" from his friend's ration.

Has your dog's activity or energy expenditure increased?

A dog nursing pups will want to consume far more calories than she usually would. Likewise, a dog that is herding sheep, running field trials, or participating in agility courses will have a bigger appetite.

Is your dog drinking more water than normal?

An increase in appetite accompanied by an increase in thirst will prompt your vet to think about an underlying hormonal imbalance.

Does your dog's vision seem normal?

Be sure to let your vet know if your dog has been bumping into things or having difficulty seeing at night. A condition called *SARDS* (sudden acquired retinal degeneration syndrome) causes increased appetite in conjunction with loss of vision.

Is your dog taking medication that might increase his appetite?

Prednisone, other steroids, and anti-seizure medications are notorious for causing an insatiable appetite.

Associated Symptoms to Report Weight gain or loss, vomiting, diarrhea, increased thirst, excessive panting, skin or hair coat changes, weakness

Symptom: Increased frequency of urination

See Urination abnormalities, p. 278

Symptom: Increased panting

Overview There's nothing quite like the feeling of lying in bed at night unable to fall asleep because your dog won't stop panting. Although you're not too warm, you check the thermostat to be sure that it's not set too high, and then begin counting sheep to get some shut-eye. You should also think about getting your pup a checkup, because all that panting *isn't* normal.

Questions Your Vet Will Ask You

Is your dog's environment warmer than usual?

The most common reason dogs pant is to dissipate body heat.

Is there anything new in your dog's environment?

Fear and anxiety—from severe weather, for example—will cause many dogs to pant.

Does your dog's body language tell you that he is in pain?

Dogs that hurt don't necessarily whimper or cry out, but they may have difficulty getting up and down or shy away when you attempt to touch a particular part of their body. Other signs are becoming reclusive, or showing reluctance to climb the stairs or jump up or down onto the furniture. In addition, they often pant excessively.

Is your dog taking any medication that could cause panting?

Prednisone and narcotics are notorious for causing this to happen.

Is your dog truly panting more, or is it simply more audible than before?

Laryngeal paralysis occurs commonly in older dogs (most often in Labrador Retrievers) and causes noisy upper respiratory sounds. Affected dogs don't necessarily pant more than usual, but the sound becomes much louder.

Associated Symptoms to Report Lethargy, fever (consider taking your dog's temperature, see p. 373), increased thirst, increased appetite, behavior change, labored breathing, change in normal gum color

Symptom: Increased thirst

Overview This symptom may appear innocuous because your dog seems to feel just fine. He's just thirstier than normal. Nonetheless, I encourage you to seek veterinary attention when you find yourself filling his water bowl more than usual. Increased thirst can be one of the earliest symptoms of a serious problem, such as kidney failure or liver disease. Note: even when he is driving you crazy with his excessive thirst (the constant refilling of the water bowl, the waking up in the middle of the night to go outside, or worse yet, the "not quite making it" outside), *never* restrict his water intake without consulting your vet.

Questions Your Vet Will Ask You

Has there been a change in diet?

Dry dog foods contain far less water content than canned foods, so expect a transition from canned to dry dog food to be accompanied by increased water intake. A food with higher salt content can also cause increased thirst.

Has there been a change in ambient temperature or activity level?

Naturally, hot weather causes increased thirst. And, when your dog's activity level is upped dramatically, expect an increase in water intake, too.

Is your dog receiving any medication that could cause increased thirst?

Prednisone, anti-seizure medication, diuretics (water pills), and several others typically increase water consumption.

Associated Symptoms to Report Change in appetite, lethargy, weight loss, vomiting, diarrhea, change in urinary habits

*** Don't Forget!**

Check p. 259 for four key questions your vet will *always* want you to answer.

Symptom: Itchy skin (pruritus)

Overview All dogs normally itch a bit. When it becomes excessive or you find yourself telling your dog to, "Stop that scratching!" it's time to set up an appointment with your vet.

Questions Your Vet Will Ask You

Is anyone else in the household itchy?

Some skin parasites—fleas are the most common—are capable of causing itchiness amongst the whole herd, humans included!

Which body parts are most affected?

When a dog is itchy from fleas, he usually chews at the area where his tail meets his rump. Inhaled allergies (a dog's version of hay fever) typically cause itchy feet and face.

Is your dog's itch seasonal or does it occur year-round?

Seasonal itching tends to occur in association with inhaled allergies (hay fever) or fleas (more problematic in the warmer months). Scratching that occurs regardless of time of year is more likely to be a food allergy.

Is there anything new in your dog's surroundings that could cause an allergic reaction?

If the itch began right at the time of the introduction to a new diet, bedding, or household cleaner, your dog may be reacting to the change. Food allergies may manifest as itchy skin following a diet change.

Did you observe a change in the appearance of your dog's skin before the itching started?

Once a dog starts biting and scratching himself, he can create all kinds of skin sores, hair loss, and inflammation. Your vet will want to know if you noticed any such changes before the scratching began.

Associated Symptoms to Report Decreased appetite, lethargy

Symptom: Labored breathing

Overview This is always a worrisome one, because many of the diseases that cause difficulty breathing require urgent attention. If you notice your dog suddenly breathing more rapidly or with greater effort, have him checked out right away.

Questions Your Vet Will Ask You

Does the labored breathing come on and end abruptly?

"Reverse sneezing" is an extremely dramatic way many dogs respond to a tickle in their throats. During a reverse sneeze, the dog hunches over, extends his head forward, and makes lots of noise when breathing in and out. The episode may last for up to a minute. It typically ends as quickly as it began. People who don't know better often interpret it to be a life-threatening asthmatic episode. The good news is that a reverse sneeze is really nothing more than a dog clearing his throat. Naturally, your dog is unlikely to demonstrate this behavior in front of your vet!

Are you hearing increased breathing noises?

If so, chances are that the problem is in the upper respiratory tract (throat or windpipe) rather than the lower respiratory tract (the lungs).

Does he breathe more easily with his mouth open rather than closed?

If so, the cause of the breathing problem is likely within your dog's nose. Try holding a small mirror up to the nose when he is breathing with his mouth closed. If air is passing normally through both sides of his nose, you should observe two small circles of condensation form on the mirror surface. If not, there is obstruction to airflow within one or both nasal passageways.

Is your dog struggling to inhale, exhale, or both?

If your dog is working hard to inhale, the problem is probably located in the throat or trachea (windpipe). If breathing out is the hard part, the origin of the problem is likely within the lungs.

Has your dog recently vomited or undergone general anesthesia?

Sometimes, during the course of normal vomiting or when it occasionally happens during anesthesia, some of the stomach contents are inhaled into the lungs, which causes an abrupt onset of labored breathing. This condition is called *aspiration pneumonia*.

Associated Symptoms to Report Weakness, cough, vomiting, decreased appetite, change in normal gum color, decreased stamina

Symptom: Lethargy

Overview A lethargic dog is one lacking in energy that has lost interest in the things that would normally excite him. People often describe him as "depressed" or "not himself." A myriad of causes for canine lethargy exist. Pay careful attention to any accompanying symptoms and try to provide your vet with the answers to the following questions.

Questions Your Vet Will Ask You

Does your dog appear lethargic around the clock or does he have periods of normalcy?

Lethargy caused by low blood sugar, liver disease, or brain disease may wax and wane.

Are there any emotional reasons your dog may be acting lethargic?

Sometimes a significant change or disruption in the household, such as the introduction of a new dog or cat, or the loss of a loved one (whether human or animal), can cause transient lethargy.

Are you giving your dog any medication that might be the cause?

Lethargy can be a side effect of many medications.

Associated Symptoms to Report Vomiting, diarrhea, changes in thirst or appetite, coughing, incoordination, decreased stamina, weight gain or loss

Symptom: Limping (lameness)

Overview Limping is caused by pain or is the result of a neurological issue within the affected leg or the back or neck. Orthopedic growth abnormalities are a common cause of lameness in young, rapidly growing dogs. Older dogs are prone to arthritis.

Questions Your Vet Will Ask You

Is the same leg always affected?

A "shifting leg lameness," rather than one that affects just one leg, will prompt your vet to eliminate a number of potential causes.

Is the lameness constant throughout the day?

Arthritis tends to be worse after rest and then improves after the dog has been up and moving about. Other causes of lameness become worse with activity.

Have you found a tender spot?

Let your veterinarian know if there is a particular spot where your dog is "ouchy."

What was your dog doing when the lameness began?

Be sure to report if your dog was doing something foolish with his body right before the limping started.

Associated Symptoms to Report Lethargy, decreased appetite, weight loss

Symptom: Lump/bump

Overview Although many are benign, all lumps and bumps deserve to be checked out. If you've previously decided on a course of benign neglect (waiting and watching) but now the mass is growing or changing in appearance, it's time to reevaluate your plan.

Questions Your Vet Will Ask You

Has your dog been bothering (itching or chewing) the mass?

Even if the mass is benign, the fact that your dog is bothered by it is a good reason to consider its removal.

Has the lump grown in size compared to when it was first noticed?

If, indeed, it is growing, convey a sense of the growth rate (i.e., doubled in size over the course of two weeks; grew by 25 percent over three month's time).

Has your dog had other lumps or bumps removed in the past? If so, what did the biopsy show?

It's not uncommon for a dog to develop the same type of tumor in more than one location over his lifetime. If a previous lump was cancerous, your vet will want to be more proactive when a new lump appears on the scene. If possible, bring along any prior biopsy reports when visiting a new vet.

Associated Symptoms to Report Lethargy, decreased appetite, weight loss, vomiting, diarrhea, coughing, labored breathing

Symptom: Nasal discharge

Overview Nasal discharge is often associated with sneezing and can be caused by disease within the sinuses or nasal passageways. Infection, allergies, tumors, nasal mites, dental disease, and foreign bodies can all be the cause.

Questions Your Vet Will Ask You

What does the discharge look like?

Is the discharge clear, yellow, green, or bloody? Your vet will also want to know if there is an odor, too, and whether the discharge is coming from one or both nostrils. By the way, don't wipe away the discharge right before the vet examines your dog. She will want to see it with her own eyes.

Is your dog sneezing more than normal? If so, did the sneezing begin abruptly?

Foreign bodies within the nasal passageway are far more likely to cause an abrupt onset of sneezing than any other cause of nasal discharge.

Does your dog run out in fields where there are weeds?

Plant material (grass and weeds in particular) is far and away the most common type of foreign body that ends up in a dog's nose.

Associated Symptoms to Report Lethargy, decreased appetite, coughing, ocular discharge, squinting, facial asymmetry, itchiness around the face

Symptom: Ocular discharge (discharge from the eyes)

Overview Common causes for eye discharge include: infection; allergies; glaucoma; inflammation within the eye (*uveitis*); a scratch on the surface of the eye (cornea); a foreign body; ingrown eyelashes; blockage of the nasolacrimal duct that normally drains tears from the eye; and decreased tear production.

Questions Your Vet Will Ask You

Is your dog squinting or rubbing at his eyes?

Some of the causes of ocular discharge also cause discomfort.

How would you describe the discharge?

Is it yellow, green, or clear? Is it affecting one or both eyes? How often do you need to clean them? (Try not to clean your dog's eyes right before your visit with your vet. She will want to see the discharge herself.)

Is your dog's nose wet or dry?

Diminished tear production often produces a dry nose in addition to the ocular discharge.

Associated Symptoms to Report Lethargy, decreased appetite, other changes in eye appearance

Symptom: Pain

Overview The causes for pain in dogs are just as numerous as for people. The big difference is that dogs can't easily convey where they hurt. If you have the perception that your dog is in pain, don't hesitate to call your vet.

Questions Your Vet Will Ask You

What gives you the impression that your dog is experiencing pain?

Not all dogs whine, whimper, or cry out; their expression of discomfort may be far more subtle. Reclusive behavior, a hunched appearance, shying away from being petted, an increased respiratory rate, and a reluctance to perform normal activities, such as climbing the stairs or stretching down to the food bowl, can all indicate a problem.

Do you have a sense of where your dog is hurting?

Is he limping or does he cry out when you touch a particular body part?

Does the pain appear in conjunction with a particular activity?

Dogs with arthritis tend to be good and sore the day after heavy exercise. A slipped disk often causes a dog to cry out when he turns his head. A dog with colitis or rectal pain may vocalize in conjunction with passing a bowel movement.

Associated Symptoms to Report Limping, tucked tail, arched back, stiffness, lethargy, restlessness, vomiting, diarrhea, coughing, sneezing, change in thirst or appetite, weight loss

Symptom: Regurgitation

See Vomiting, p. 280

Symptom: "Scooting"

Overview What is "scooting," you might ask? It is that rather embarrassing thing your dog does to itch the area around his anus—you know, when he sits on the ground and then drags himself forward with his front legs. Why your dog always chooses to do this when you have company visiting is anyone's guess (that's when kitties like to vomit as well). Perhaps they're trying to get the guests to leave sooner so that they can get to the leftovers!

Questions Your Vet Will Ask You

Have you seen any evidence of fleas?

This is important for two reasons. Firstly, fleas cause itchy skin that might cause your dog to scoot. Secondly, fleas transmit tapeworms. When, in the course of grooming himself, your dog happens to ingest a flea, he might develop tapeworms—long, segmented, white parasites that live in a dog's small intestine. No, they do not rob the body of its nutrition as the old wives' tale would have you believe. While in rare cases there might be so many tapeworms present they could cause a partial bowel obstruction, 99 percent of the time they are merely an aesthetic issue—the adult worms release small segments that migrate out the dog's anus, resulting in the desire to scratch and scoot.

Have you observed any dried tapeworm segments around the anal region?

After the tapeworm segments migrate out the anus they quickly dry out, looking like little sesame seeds.

Have you observed any blood in the stool or an unusual odor around your dog's hind end?

Every dog comes equipped with two anal sacs that sit beneath the skin surface at the ten and two o'clock positions of the anus. Most dogs empty the secretions from these sacs in conjunction with having a bowel movement. Sometimes the material in the anal sac becomes infected, and bloody discharge or an inappropriate release of the anal sac contents (smells about as bad as anything you could imagine smelling) is the result. When a dog has anal sac disease, their natural tendency is to want to scoot! Your vet will empty the anal sacs and prescribe antibiotics if indicated. She may also suggest having them expressed (emptied) on a regular basis.

Associated Symptoms to Report Changes in stool appearance or bowel behavior, abnormal appearance of the anus or surrounding skin

> **❊ Don't Forget!**
> Check p. 259 for four key questions your vet will *always* want you to answer.

Symptom: Seizure

Overview If you have just watched your dog's very first seizure, you are, no doubt, feeling frightened and traumatized—and helpless, as well. Do your best not to panic. I realize this would have been like asking me to remain calm during my first California earthquake! Believe it or not, once you have observed a number of seizures, they no longer seem nearly as scary. Most of the time, a seizure is short-lived, and the dog is back to normal within minutes. The likelihood of a single seizure causing a sustained problem or death is slim.

When your dog has a seizure, take a few deep breaths and concentrate on making observations. If he is in a precarious location, try to block his body to prevent him from falling. *Do not*, under any circumstances, put your hand in his mouth while he is having a seizure. The notion that he will swallow his tongue is a myth; sustaining a serious bite wound while trying to grab hold of it is not! Once the seizure has concluded, call your vet for advice on how to proceed. If the seizure lasts longer than two minutes, make preparations to take your dog to the nearest open veterinary facility at once.

Questions Your Vet Will Ask You

Have there been prior seizures?

If your dog has had a seizure before—even years ago—your veterinarian will more strongly consider the diagnosis of *epilepsy*, which is a seizure disorder of undetermined cause.

Was your dog conscious during the episode?

Sometimes, it can be difficult to tell if a dog is having a seizure or a fainting episode. Seizures typically cause the dog to become unconscious (complete lack of responsiveness). In contrast, a dog that has fainted usually responds to loud noises or familiar human voices. Seizures and fainting have entirely different causes, so noting the difference can be truly helpful information for your vet.

Did your dog drool or lose control of urine or bowels during the episode?

If any of these occurred, what you saw was likely a seizure rather than fainting. They may also serve as the only evidence of a seizure that occurred when you were not home.

What did the seizure look like?

Seizures are often generalized, meaning the dog is lying down and experiencing involuntary muscle activity. Sometimes a seizure will manifest as a dazed look: the lights are on, so to speak, but no one is home. This information can sometimes help track down the underlying cause.

What time of day did the seizure occur?

Seizures can be caused by a low blood-sugar concentration. When this is the case, the seizure is more likely to take place right before mealtime (when blood sugar is lowest) rather than following a meal. On the other hand, if liver disease is the cause, a seizure is more likely to happen immediately following a meal.

Has your dog been exposed to anything new or different during the last couple of days?

Some toxins (antifreeze, poisonous mushrooms, chocolate, snail bait), medication, and chemical substances found in the home or yard can cause seizures. It is worthwhile mentioning anything new that your dog may have been in contact with. If possible, canvass the surroundings for any potential culprits.

Associated Symptoms to Report Behavior changes, lethargy, incoordination, loss of vision, change in appetite or thirst, vomiting, weight loss

Symptom: Sneezing

Overview All dogs sneeze occasionally and some do so habitually when excited. This is normal. Causes for *excessive* sneezing include: infection; allergies; foreign bodies; trauma; and tumors within the nasal cavity or sinuses.

Questions Your Vet Will Ask You

Does your dog run in fields where there are weeds?

The nasal passageways are common places for plant foreign bodies to lodge.

Did anything in your dog's environment change around the time of the onset of sneezing?

If someone in your home took up smoking, or you've recently used a new carpet cleaner, the sneezing may be an allergic response.

Is nasal discharge present?

If so, be prepared to describe its appearance to your veterinarian (see Nasal discharge, p. 274).

Associated Symptoms to Report Coughing, decreased appetite, lethargy, weight loss

Symptom: Straining to have a bowel movement

See Diarrhea, p. 266

Symptom: Straining to urinate

See Urination abnormalities, below

Symptom: Urination abnormalities

Overview Several abnormalities associated with urination occur in dogs, including: involuntary urination (incontinence); straining to urinate; increased frequency of urination; inappropriate urination (in the house, for example); decreased size of urine stream; and abnormal urine odor or color. Any of these symptoms should be reported. An inability to urinate (straining with nothing coming out) requires urgent veterinary attention. Potential causes of urination abnormalities include: infection; urinary tract stones; structural defects and tumors. Keep in mind that because of their anatomical connections, diseases within reproductive structures (prostate gland, uterus, vagina) are capable of causing urinary tract symptoms.

Questions Your Vet Will Ask You

Is your dog drinking more water than normal?

Increased thirst naturally results in increased urine volume. This might explain why your pup is waking you up in the middle of the night to urinate or experiencing urinary incontinence (urine leakage).

If blood is present in the urine, is it apparent throughout the urine stream or only at the end of urination?

Blood is heavier than urine, so it tends to "sink to the bottom" of the bladder. And, when blood pools there, blood seen in the urine will steadily become more apparent as the bladder empties. If blood is present to the same degree from start to finish, a problem in the urethra (the tube that leads from the bladder to the outside world), prostate gland (male), or vagina (female) is more suspect.

If your dog is urinating in the house, are you finding small puddles or large puddles?

Most dogs with urinary tract infections feel the need to urinate even when only a small amount of urine has accumulated within the bladder. They tend to leave small rather than large puddles. Large puddles imply bladder overdistention caused by increased thirst and water consumption or involuntary urine leakage.

Are you finding urine puddles where your dog normally sleeps?

Dogs tend to avoid soiling near their bed, so this may indicate he is involuntarily leaking urine (urinary incontinence) rather than actively urinating.

Have you observed discharge or blood dripping independently of urination?

When this happens, your vet will know that the bleeding is originating from the prostate gland (male), vagina or uterus (female), or the urethra.

Has your dog been straining to have a bowel movement?

The lower urinary tract is in close proximity to the rectum. Diseases that affect one are capable of affecting the other.

Associated Symptoms to Report Increased thirst, lethargy, change in appetite, vomiting, diarrhea, weight loss

Symptom: Urine leakage (urinary incontinence)

See Urination abnormalities, p. 278

Symptom: Vaginal discharge

Overview Vaginal discharge readily goes unnoticed in a longhaired dog. If you smell a funny odor or male dogs have suddenly become interested in your darling girl, consciously take a look at this part of her anatomy.

Questions Your Vet Will Ask You

Has your dog been spayed?

If your dog has not been spayed, familiarize yourself with her normal heat cycle. Most intact females will have vaginal discharge for one to three weeks approximately every six months. This is the time when they are fertile.

Has your dog been bred?

Some sexually transmitted diseases can cause a vaginal discharge. There are a number of such discharges that are associated with pregnancy and giving birth.

If your dog has not been spayed, when was she last in heat?

Pyometra (an accumulation of pus within the uterus—see p. 350) tends to occur within two to six weeks after a dog has been in heat. If the dog has an "open pyometra" (the cervix is open), vaginal discharge will be apparent. There is no vaginal discharge when the pyometra is associated with a closed cervix.

What is the nature of the discharge?

Is the discharge clear or bloody? Does it resemble mucous or pus? Is there an associated odor? Is it constant or intermittent?

Has the discharge been present in the past? If so, how long ago?

During a normal spay procedure, the uterus and both ovaries are removed. Occasionally, a tiny bit of ovarian tissue is inadvertently left behind and this is capable of inducing ongoing heat cycles (vaginal discharge approximately every six months).

Have you observed any changes in your dog's urination?

The urinary tract connects with the vagina. Diseases within the lower part of the urinary tract can produce vaginal discharge along with other symptoms, such as urine leakage, "accidents" in the house, increased frequency of urination, straining to urinate, or decreased size of the urine stream.

Does your dog have exposure to foxtails?

Foxtails are weeds that grow in the western US. If you live in "foxtail country," know that this plant awn has a propensity for getting lodged within the vagina, sometimes resulting in discomfort and discharge.

Associated Symptoms to Report Lethargy, decreased appetite, vomiting, diarrhea, increased thirst, change in urinary habits

Symptom: Vomiting/regurgitation

Overview It is natural to think that vomiting is caused by disease within the stomach or intestines. This isn't always the case. Vomiting can result from a hormonal imbalance called *Addison's disease* (p. 298), as well as disease within the liver, pancreas, kidneys, uterus, and central nervous system.

It is important to distinguish whether your dog is *vomiting* or *regurgitating*. Here is the difference: Vomiting originates from the stomach or small intestine and is preceded by retching (the dog is well aware of the fact that he is about to vomit). Regurgitation originates from the esophagus (the muscular tube that carries food and water from the mouth down into the stomach) or stomach and tends to occur *without* warning. In fact, the dog is usually as surprised by the act as much as the human who witnesses it. This is why aspiration of some of the regurgitated material into the lungs (*aspiration pneu-*

monia) occurs so commonly. Vomiting and regurgitation call for altogether different diagnostic testing, so it is important to distinguish between the two.

Questions Your Vet Will Ask You

Has there been a diet change?

If the vomiting or regurgitation began coincidentally with a diet change, this transition may be the cause.

Has your dog eaten anything unusual?

Accessing the garbage can or compost pile (dietary indiscretion) is a common cause of vomiting. Well-intentioned people sometimes get carried away when feeding table scraps (all the fat you cut off your steak isn't good for your dog either)—another common cause of vomiting. Think about whether your dog ingested something that might have become lodged within the esophagus, stomach, or intestines. All kinds of items, such as bones, dog toys, children's toys, pantyhose, underwear (you name it, we've seen it) are capable of getting hung up in the GI tract.

How frequently is the vomiting or regurgitation occurring?

If the frequency waxes and wanes, report the *most frequent occurrences* (i.e., three times a day) as well as the longest your dog has been able to go *without* vomiting or regurgitating (i.e., two weeks).

What does the vomited or regurgitated material look like?

Even if the appearance varies, report what you most commonly see. If your dog brings up food, does it appear to be digested or undigested? If there is a fluid component, let your vet know if it is clear, brown, green, or yellow. Be sure to report if blood is present.

Does the vomiting or regurgitation seem related to eating or drinking?

Is there a timeframe after eating or drinking when your dog vomits or regurgitates? Does it happen after every meal? Does it only happen after he eats grass?

Does the consistency of your dog's food have an impact on whether or not vomiting or regurgitation occurs?

A structural abnormality or gastrointestinal motility disturbance may allow one particular texture or thickness of food to pass more readily than another.

Associated Symptoms to Report Lethargy, decreased appetite, diarrhea, change in thirst, weight loss, nasal discharge, coughing

✻ Don't Forget!

Check p. 259 for four key questions your vet will *always* want you to answer.

Symptom: Weight loss

Overview It is usually just as difficult for dogs to lose weight as it is for us. Although you might be thrilled when your 80-pound Boxer (who should weigh 60 pounds) drops weight, if you have not substantially reduced his caloric intake, this loss is cause for concern.

Questions Your Vet Will Ask You

Has your dog been eating normally?

An increased or normal appetite accompanied by weight loss can occur with diabetes mellitus, gastrointestinal disease, certain types of cancer, or oversupplementation of thyroid medication. If your dog's appetite has decreased, it is important to look for an underlying cause.

Has there been a change in dog food?

Simply by changing your dog's brand or type of food, you may have inadvertently decreased his caloric intake. For example, a cup of canned food typically contains twice the number of calories as a cup of dry food. A diet formulated for a senior citizen might be 10 to 20 percent less fattening than a regular adult formula.

Has your dog's energy expenditure changed?

If it's hunting or breeding season, or you and your best buddy have taken up running 5 miles a day, he's going to need more calories. A dog that is nursing pups might require three to four times more calories than normal. So, if your dog's caloric intake has not kept pace with the number of calories he's burning, this might explain the weight loss.

Is there any new competition at the food bowl?

If a new pup has come into your life, or the current canine pecking order seems to be changing, the dog that is losing weight may be missing out at mealtime.

Associated Symptoms to Report Increased thirst, vomiting, diarrhea, coughing, sneezing, increased or decreased activity level, change in behavior

Appendix II

Common Diseases and Questions You Should Ask Your Vet about Them

In the text to follow I have provided brief descriptions of common canine diseases. Each description includes the cause (if known), associated symptoms, means of diagnosis and treatment, and—most importantly—a list of key questions to ask your veterinarian. I have grouped the diseases by body system and, beginning on p. 284, listed them alphabetically within the following categories:

I encourage you to add to the list of Questions for Your Vet based on issues that are unique for you and your dog. In addition, remember to always include the Universal Questions I described in chapter 7 on p. 135:

1. **What's the prognosis?**
2. **What should I be watching for?**
3. **Is there anything I should be doing differently at home for my dog?**
4. **What should I do in case of an emergency?**
5. **What happens next?**

In addition, I urge you to refer to the general questions found on pp. 139–163, especially if the discussion involves general anesthesia (p. 148), surgery (p. 152), or special procedures (p. 146). (Note: if your dog has been diagnosed with cancer, please refer to chapter 8—it contains all the information you need to prepare a specialized list of questions to ask your veterinarian.)

Please listen carefully to what your veterinarian has to say before asking your questions. You may just find that she's managed to answer each and every question on your list!

Cardiovascular Diseases

Introduction to the Cardiovascular System

The cardiovascular system includes the heart, the major vessels that take blood to and from the heart, and the many arteries and veins that travel throughout the body. *Arteries*, including the aorta, *transport blood away from the heart* into the tissues. *Veins*, including the cranial and caudal vena cavae, *return blood to the heart*. When a blood sample is needed for analysis, it is commonly taken from the jugular vein, located in the neck.

Note: Unlike people, dogs don't live long enough to experience "clogging of the arteries" (coronary artery disease)—therefore, they don't suffer "heart attacks" as we understand them. Instead, when heart-related death occurs in dogs, it is usually the result of a severe rhythm disturbance that interferes with normal heart function, or congestive heart failure. In the latter situation—which I often refer to in this section—fluid accumulates in or around the lungs, or in the abdominal cavity as a result of "pump failure" (see more about this condition on p. 291).

QUICK REFERENCE

Which Specialist Is Right for My Dog?

Board certified veterinary cardiologists are the specialists best equipped to work with cardiovascular problems that are challenging to diagnose or treat. If a cardiologist is not available in your area, referral may be made to a specialist in internal medicine.

QUICK REFERENCE

Alphabetical Index of Featured Diseases

Alphabetical Index of Featured Diseases cont.

Diagnosis: Arrhythmia (cardiac rhythm abnormality)

The heart normally beats with a continuous steady rhythm that produces a regular appearing electrocardiogram (ECG) tracing. An abnormal heart rate or irregular rhythm is referred to as an *arrhythmia.* Mild arrhythmias produce no symptoms whatsoever. More severe rhythm disturbances can cause lethargy, weakness, decreased stamina, syncopal episodes (fainting), and even sudden death. Boxers and Miniature Schnauzers are particularly predisposed to arrhythmias. Heart disease, anemia, abnormal potassium or calcium blood levels, and diseases in other organs that "irritate" the heart's electrical system are the most common causes.

Heart rhythm disturbances are diagnosed via an ECG. Medication to control the rhythm abnormality may be recommended and, whenever possible, the underlying cause of the arrhythmia should be addressed. Prognosis is dependent on the cause and severity of the arrhythmia.

Questions for Your Vet

1. **Do you know what is causing my dog's arrhythmia? If so, how should it be treated? If not, what can we do to figure it out?**
2. **Is the arrhythmia harmful? Does it need to be treated at this time?**
3. **Does his activity level need to be modified?**
4. **Is there a way for me to monitor my dog's heart rhythm at home?**
5. **What symptoms should prompt an emergency visit?**

Diagnosis: Boxer cardiomyopathy (arrhythmogenic right ventricular cardiomyopathy)

Arrhythmogenic right ventricular cardiomyopathy (ARVC) is a disease of the heart muscle that affects some Boxers. Only rarely are other breeds affected. The electrical activity in the heart that causes synchronous contractions becomes disrupted resulting in an abnormal heart rhythm. When mild, dogs show no outward symptoms. More severely affected dogs experience syncopal episodes (fainting) or even sudden death when the arrhythmia interferes with normal cardiac output. Boxer cardiomyopathy occasionally results in congestive heart failure (see p. 291). The best predictor of whether or not a Boxer will be afflicted is family history.

QUICK REFERENCE

Breed-Related Heart Disease

To find out what type of heart disease your favorite breed is predisposed to, talk to your veterinarian and visit the congenital heart disease registry at Orthopedic Foundation for Animals (www.offa.org/cardiacinfo.html) as well as reputable breed-specific Web sites.

Diagnosis is made by an electrocardiogram (ECG). A 24-hour or event-triggered ECG monitoring equipment can be used when the arrhythmia occurs only sporadically. There is no cure for Boxer car-

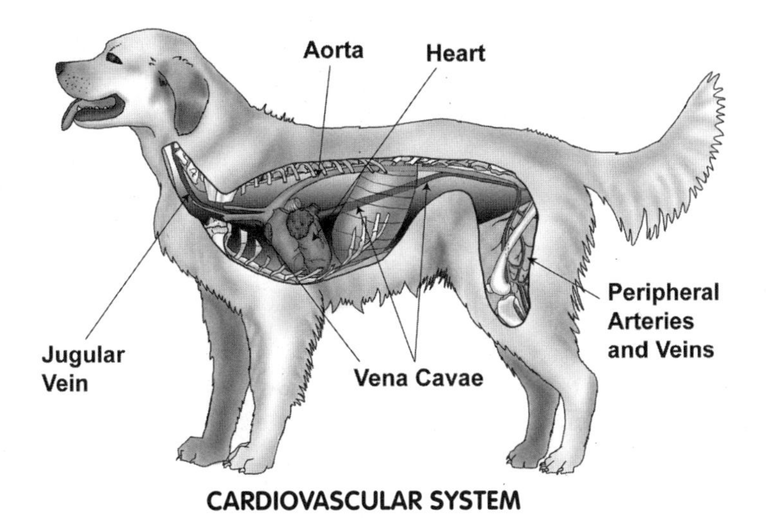

CARDIOVASCULAR SYSTEM

diomyopathy. The disease is treated with medication to control the heart rhythm abnormality. Affected dogs should not be bred.

Questions for Your Vet

1. **Should I have my Boxer periodically screened for cardiomyopathy even if he is showing no symptoms? If so, how often?**
2. **How severe is his arrhythmia? Is it likely we can control it with medication?**
3. **Can we reduce or eliminate his fainting episodes?**
4. **How frequently will you want to monitor my dog's ECG?**
5. **Should I alter his normal activity level?**
6. **What symptoms should prompt an emergency visit?**

Diagnosis: Congenital heart disease

A congenital disease is one that is present from birth. *Congenital heart diseases* include defects in the normal structure and function of the heart wall, valves, or major blood vessels. Certain breeds are associated with specific congenital defects; for example, Labrador Retrievers are prone to malformation of the tricuspid valve (*tricuspid valve dysplasia*), while *aortic stenosis* (narrowing of the major artery that carries blood from the heart) occurs more commonly in Golden Retrievers, Newfoundlands, and German Shepherds.

Some defects never cause a problem while others significantly impact heart function, quality of life, and longevity. Usually, the first sign of a congenital heart defect is a heart murmur (abnormal sound) heard by the vet during a "well-puppy exam." Identification of the type and seriousness of the defect can be done with an echocardiogram (ultrasound evaluation of the heart). Note: Interpreting the echocardiogram can be a tricky business—the more experience the ultrasonographer has, the better. Some congenital heart defects can result in congestive heart failure. X-rays of the heart and lungs help determine if this is the case.

Symptoms vary with the type and severity of the heart defect and can include: decreased stamina; labored breathing; purple gums; collapse; and even sudden death. Some defects can be corrected surgically; others are treated medically. Some require no treatment, and for a few, there is no treatment available.

Questions for Your Vet

1. **What type of congenital abnormality does my dog have? How serious is it?**
2. **What symptoms do you anticipate and over what time period?**
3. **How frequently will you want to recheck my dog's heart?**
4. **Should his activity level be restricted?**
5. **What symptoms should prompt an emergency visit?**

Diagnosis: Dilated cardiomyopathy

Dilated cardiomyopathy (DCM) is a disease of the heart muscle in which contractions are profoundly weakened and normal blood circulation is impaired. The heart enlarges with thin, flabby walls. Genetics likely play a role in the development of this disease. Doberman Pinchers, Irish Wolfhounds, Newfoundlands, Scottish Deerhounds, and Cocker Spaniels are some of the more commonly affected breeds. Occasionally, cardiomyopathy occurs as a result of a deficiency in an amino acid (protein building block) called *taurine*. Although some dogs with dilated cardiomyopathy remain free of symptoms, over time most develop severe heart rhythm abnormalities and/or congestive heart failure.

Symptoms can include: weakness; labored breathing; cough; decreased appetite; lethargy; weight loss; and abdominal distention. The diagnosis of dilated cardiomyopathy is confirmed via an echocardiogram (ultrasound of the heart). Treatment consists of medication to: control excessive fluid accumulation; increase the strength of cardiac contractions; maintain a normal heart rhythm; control blood pressure; and reduce the workload on the heart. Decreased exercise, salt restriction, and occasionally taurine supplementation are recommended. For dogs in critical condition with congestive heart failure, hospitalization is sometimes necessary to monitor them and give supplemental oxygen therapy and intravenous medication. The long-term prognosis is poor; the short-term prognosis is variable and depends on response to treatment.

Questions for Your Vet

1. **If my dog is a breed at risk, should he be screened for the disease even if he is free of symptoms?**

If so, at what age and how often?

2. **How serious is my dog's cardiomyopathy? Is he close to or even in congestive heart failure?**
3. **Is there a way for me to monitor his heart rate and rhythm, as well as his respiratory rate at home?**
4. **Should his activity level be restricted?**
5. **Should his diet be altered?**
6. **How frequently should my dog be reevaluated?**
7. **What symptoms should prompt an emergency visit?**

❊ **Don't Forget!** Check p. 283 for Universal Questions, and pp. 148 and 152 for questions when dealing with anesthesia and surgery.

Diagnosis: Endocarditis (bacterial endocarditis, infective endocarditis)

Endocarditis is a bacterial infection of the inside lining of the heart (the *endocardium*). It causes the formation of a vegetative lesion (clump of bacterial/inflammatory debris) on a heart valve, disrupting its normal function and showering bacteria and inflammatory debris into the bloodstream. Symptoms include: fever; lethargy; weakness; decreased appetite; and intermittent lameness. If the valvular dysfunction is serious enough congestive heart failure may result (see below).

The diagnosis is based on characteristic symptoms, the presence of a new heart murmur, and finding the heart valve abnormality with ultrasound. Blood cultures identify the causative bacteria. Long-term antibiotics are the cornerstone of therapy along with treatment of congestive heart failure, if present. The prognosis is dependent on the degree of heart valve damage.

Questions for Your Vet

1. **What heart valve is affected?**
2. **How serious is the endocarditis? Is my dog in or near congestive heart failure?**
3. **Should he be treated in the hospital or at home?**
4. **Should his activity level be restricted?**
5. **How frequently should he be reevaluated to see if congestive heart failure is occurring?**
6. **What symptoms should prompt an emergency visit?**

Diagnosis: Heart failure (congestive heart failure)

The heart pumps blood around the body at a rate and with the force necessary for adequate tissue oxygenation. *Congestive heart failure* occurs when the heart's pumping activity can no longer keep up with oxygen demands. Although "heart failure" describes the dog's status, it doesn't indicate the underlying cause of that failure. Causes may include: heart muscle disease (cardiomyopathy); valvular disease; congenital heart defects; arrhythmias; heartworm disease; and some types of chronic lung disease.

Symptoms of heart failure can include: decreased stamina; cough; labored breathing; abdominal distention; weakness; syncopal episodes (fainting); and sudden death. Fluid may accumulate in the lungs, chest cavity, and abdominal cavity. Diagnosis is confirmed with chest X-rays and an echocardiogram (ultrasound evaluation of the heart). Blood tests determine the impact of congestion and decreased blood flow on the liver and kidneys.

Treatment of congestive heart failure typically includes medication to: mobilize accumulated fluid from the body; decrease the workload on the heart; increase the heart's strength of contractions; and control heart rhythm abnormalities. Supplemental oxygen therapy is needed for some dogs until their heart failure crisis has resolved. Rarely can congestive heart failure be cured, but it can often be successfully managed with medication. The prognosis depends on the underlying heart disease and response to therapy.

Questions for Your Vet

1. **Do you know what is causing my dog's heart failure? If so, what can be done to treat it? If not, what can be done to determine the cause?**
2. **How severe is his heart failure? Is he in need of supplemental oxygen therapy?**
3. **Have any other organs been damaged as a result?**
4. **Should his diet be altered to minimize salt intake?**
5. **Should his activity level be restricted?**
6. **Should I be monitoring any parameters at home (gum color, heart and respiratory rates)?**
7. **What symptoms warrant an emergency hospital visit?**

Diagnosis: Heartworm disease

Heartworms *(Dirofilaria immitis)* are long, white, spaghetti-like worms that live in the right side of the heart, as well as the pulmonary arteries (blood vessels that carry blood from the right heart into the lungs). Mosquitoes transmit *heartworm disease*: they ingest *microfilaria* (immature heartworms) while feeding on an infected dog, and then inoculate the microfilaria into the next unlucky dog that supplies a meal. Some heartworm infections are "sterile"—no microfilaria are found within the bloodstream. But, even dogs with "sterile infections" develop heartworm symptoms due to the presence of adult worms—they just aren't "carriers" that pass infection to other dogs via mosquitoes.

Heartworm disease most commonly causes symptoms associated with an allergic reaction to the worms, and eventually, congestive heart failure. Symptoms tend to progress gradually over time and most commonly include: decreased stamina; coughing; labored breathing; weight loss; and syncopal episodes (fainting). Some dogs with heartworm disease have secondary kidney damage that causes excessive protein leakage into the urine (see Glomerular disease, p. 368).

Heartworm disease is diagnosed by simple blood tests to document the presence of adult worms and microfilaria. The three stages of heartworm treatment are: 1) adulticide therapy (elimination of the adult worms with two or more injections of medication); 2) microfilaricide therapy (elimination of the immature worms in the bloodstream with oral or injectable medication); and 3) initiation of heartworm preventative therapy.

Following adulticide treatment, strict confinement (no running, jumping, playing) is mandatory. This helps prevent blockage of the lungs' blood vessels by dying worms that are released from the heart into circulation—a complication more likely in severe cases. The dose of adulticide is often divided and given over two months rather than all at once in order to achieve a more gradual die-off of the heartworms and hopefully prevent this life-threatening complication.

Questions for Your Vet

1. **How severe is my dog's heartworm disease?**
2. **Have his lungs or kidneys been affected? If so, how should they be treated?**
3. **Do you recommend splitting the adulticide treatment over two months?**
4. **How strict must the post-adulticide confinement be and for how long? What is the best way to keep my dog confined?**
5. **Does my dog have circulating microfilaria? If so, when should medication be given to treat them?**
6. **Should my other dogs be tested for heartworm disease?**
7. **When should my dog be retested for heartworm disease?**
8. **What heartworm preventative do you recommend? When should it be started? Should it be given year round?**

Diagnosis: Mitral/tricuspid valve disease (myxomatous degeneration, endocardiosis)

Mitral/tricuspid valve disease is an age-related degenerative process of the heart valves that causes them to become thickened and leaky. The mitral valve, which separates the left atrium from the left ventricle, and/or the tricuspid valve, which separates the right atrium from the right ventricle, are affected. When the heart contracts, affected valves allow reverse regurgitant blood flow from the ventricles back into the atria (normally the valve would be sealed tight). This increases the workload upon the heart. If the valve disease is severe and chronic enough, it can cause congestive heart failure.

Mitral/tricuspid valve disease is common in small breed dogs. One of its symptoms is a heart murmur that can be heard with a stethoscope, and sometimes even felt on the chest wall. Murmurs are graded on a scale of "I" to "VI" based on their loudness for purposes of assessing change over time. A change in the grade of the heart murmur might warrant more detailed investigation. However, keep in mind that the grade of the heart murmur does not always correlate with the severity of the disease.

Diagnosis is confirmed with an echocardiogram (an ultrasound evaluation of the heart). Many dogs with this disease remain unaffected throughout their lives. For others, the valve changes ultimately lead to congestive heart failure (see p. 291).

Questions for Your Vet

1. **Is this the first time a murmur has been heard?**
2. **What grade of heart murmur is it? How does this compare to any heart murmur previously heard?**
3. **Is my dog currently in a state of congestive heart failure? If not, is any treatment recommended at this time?**
4. **Should my dog's activity level be restricted?**
5. **How frequently should his heart be reevaluated?**

Diagnosis: Pericardial effusion (pericardial fluid)

The heart sits within a thin impermeable membrane called the *pericardial sac*. Normally, this sac is devoid of fluid. When fluid accumulates within the pericardial sac, it acts like a pressure bandage preventing the heart from filling with blood and contracting normally. Resulting symptoms include: weakness; collapse (inability to rise); increased heart and respiratory rates; and distention of the abdomen with fluid.

When the fluid within the pericardial sac is blood, the most common causes are: bleeding from a tumor attached to the heart; idiopathic pericardial effusion (no apparent source of the bleeding can be found); and blood clotting abnormalities. Potential causes of fluid other than blood within the pericardial sac are heart failure and infection.

Ultrasound of the heart (echocardiogram) and collection of a pericardial fluid sample for analysis (done via insertion of a small needle into the pericardial space) are used to document the presence of pericardial effusion and determine the cause. Drainage of the fluid may be necessary on an emergency basis. Chronic therapy involves treatment of the underlying disease and, when needed, drainage of the pericardial fluid. Surgery to create a permanent "window" in the pericardial sac to allow for constant drainage is sometimes considered. The prognosis is variable with idiopathic pericardial effusion, but poor with a heart-based tumor.

Questions for Your Vet

1. **Do we know what is causing the pericardial effusion? If so, how can it be treated? If not, what can be done to determine the cause?**
2. **Is there value in draining the fluid? If so what are the associated risks?**
3. **Is it likely that the fluid will recur?**
4. **How far should I allow symptoms to progress before bringing my dog in for reevaluation?**
5. **What symptoms should prompt an emergency visit?**

Dental Diseases

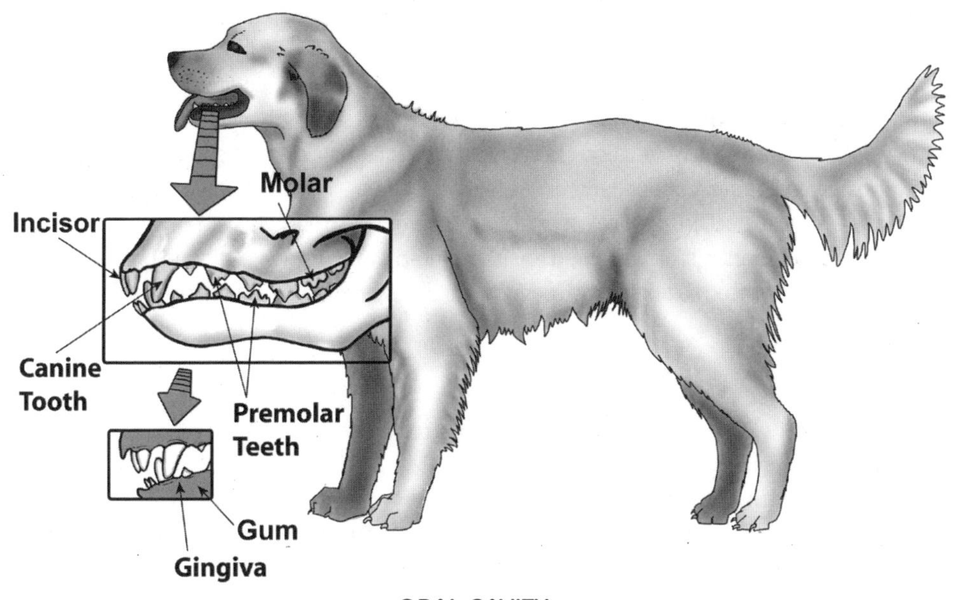

ORAL CAVITY

Introduction to the Oral Cavity

Dogs begin life with no teeth apparent. By the time they are eight weeks old, they have 28 deciduous (baby) teeth, which they lose at approximately four months of age. The mouth then fills with 42 teeth consisting of 12 incisors, four canine teeth, 16 premolars, and 10 molars. The teeth are secured in place by the jaw bones and surrounding gingival (gum) tissue.

QUICK REFERENCE

Which Specialist is Right for My Dog?

Most dental work is performed in the general practice setting. However, some advanced dental procedures require specialized equipment and training so referral to a board certified veterinary dentist may be recommended.

Diagnosis: Periodontal disease (gingivitis)

Periodontal disease refers to inflammation and infection of the gums and other structures that support the teeth. It is caused by plaque and tartar accumulation that allow bacteria to grow along the gum line. Eventually pockets form around the teeth under the gum line—the perfect haven for bacteria to flourish. Although soft-food diets predispose to periodontal disease, genetics play a major role. Small breed dogs are particularly prone to periodontal disease.

Symptoms include: red and swollen gums with discharge and/or bleeding; loose teeth; consequent pain; and halitosis (bad breath). Diagnosis of periodontal disease is based on examination of the oral cavity (mouth). Dental X-rays are used to evaluate the tooth roots and the supporting bone. Treatment involves dental cleaning under general anesthesia. Deep pockets are treated surgically. Infected teeth are treated with extraction (removal) or *endodontic surgery* (a "root canal"—removal of the tooth's nerve tissue and blood vessels). Afterward, a routine home-prevention plan is mandatory or periodontal disease will quickly recur. Left untreated, periodontal disease can be a source of infection for other sites in the body.

Questions for Your Vet

1. **How severe is my dog's dental disease? Are there any infected teeth?**
2. **Do you sense that my dog is in pain?**
3. **Is there need for antibiotics before, during, or following the dental work?**
4. **Which at-home hygiene/prevention program is likely to be the most successful?**
5. **What things are safe for my dog to chew on?**

❊ **Don't Forget!** Check p. 283 for Universal Questions, and pp. 148 and 152 for questions when dealing with anesthesia and surgery.

Diagnosis: Tooth fracture (broken tooth)

Fractured teeth can be caused by chewing on hard objects, dog fights, and other trauma. Fractures are diagnosed by visual inspection and dental X-rays. *Superficial* or *uncomplicated* fractures involve only the outer layers of the tooth. *Complicated* fractures extend into the tooth pulp where the nerves and blood vessels that supply the tooth are located.

The treatment plan is determined by whether a fracture is uncomplicated or complicated. An uncomplicated fracture can often be left untreated, but with ongoing monitoring. To prevent pain and infection, complicated fractures are treated with *vital pulp therapy* (a procedure performed when the fracture is recent in the hopes of salvaging remaining healthy nerve tissue and blood vessels within the tooth), *endodontic surgery* (see above), or removal of the tooth.

Questions for Your Vet

1. **Can you show me where the tooth is broken?**
2. **Is the fracture complicated or uncomplicated? If uncomplicated, how frequently should we monitor the appearance of the tooth?**
3. **Is there any other oral cavity trauma?**

❹ **Do you sense that my dog is in pain?**

Diagnosis: Tooth root infection (apical abscess)

Just like people, dogs can get infections in the roots of the teeth. The first symptom is often an awful odor. Surprisingly, many dogs show no overt evidence of pain, but others start to prefer soft foods over hard, chew only on the unaffected side, or approach the food bowl with enthusiasm only to turn away. Over time, the crown of the tooth becomes discolored and the tooth becomes loose. Dental X-rays confirm the diagnosis. Treatment options are *endodontic surgery* (see p. 296) or removal of the tooth.

Questions for Your Vet

❶ **Do you think my dog is in pain?**
❷ **Is there a need for antibiotics?**
❸ **What can I do to prevent other teeth from becoming infected?**
❹ **Should I change his diet to a softer food?**

Endocrine Diseases

Introduction to the Endocrine System

Hormones are chemical substances produced within glands and released into the bloodstream in order to physiologically impact a cell or organ at a distant site. Endocrine (hormonal) diseases are caused by the under- or overproduction of a particular hormone and, in the dog, are most commonly associated with abnormalities of the pancreas, pituitary, adrenal, thyroid and parathyroid glands. Diseases associated with the ovaries and testicles (major hormone producers) are discussed in the section on reproductive diseases (see p. 345).

QUICK REFERENCE

Which Specialist Is Right for My Dog?

Diagnosing, and then regulating a hormone imbalance can be challenging. If help is needed, you and your veterinarian should enlist the assistance of a board certified veterinary internist.

Diagnosis: Addison's disease (hypoadrenocorticism)

It may sound strange, but as a veterinary internist, there are some diseases I am actually happy to diagnose. *Addison's disease* is a favorite of mine because it's one of the few I see that is readily treatable and has an excellent prognosis. The downside is that it is costly to treat: the medications are pricey—the bigger the dog, the more expensive the drug—and treatment is lifelong.

Every dog has two adrenal glands. Addison's disease occurs when these glands quit producing the hormones *cortisone* and *aldosterone*. Cortisone is a "fight-or-flight" hormone, essential for the body to deal with normal day-to-day stress. Aldosterone is the hormone responsible for maintaining normal blood levels of sodium and potassium.

Dogs with Addison's disease tend to have waxing and waning symptoms such as: weakness; malaise; poor appetite; vomiting; diarrhea; increased thirst; and abnormal levels of sodium and potassium in the bloodstream. The disease can be effectively managed with daily oral medication or by an injection given once every 25 days, approximately. Dosage is adjusted as needed to resolve symptoms and restore normal energy and sodium and potassium levels. With careful treatment and monitoring the prognosis is excellent.

Questions for Your Vet

- ❶ **What are the pros and cons of the treatment options?**
- ❷ **If I choose to treat with the monthly injection, can I be trained to do this at home?**
- ❹ **Should I begin salting my dog's food to provide extra sodium?**
- ❺ **Will my dog need supplemental cortisone for stressful events or activities (traveling, boarding, visiting the groomer, going to the dog park)?**

Diagnosis: Cushing's syndrome (hyperadrenocorticism)

In some middle-aged and older dogs one or both adrenal glands become overactive, producing too much cortisone and causing a disease we call *Cushing's syndrome*. Cushingoid symptoms can include: increased thirst (and urination); increased appetite (the perfectly behaved dog starts to raid the garbage can); panting; muscle weakness resulting in a pot-bellied appearance; hair loss; and susceptibility to skin and bladder infections. These symptoms can take years to fully develop and vary from dog to dog. High blood pressure and excessive protein leakage into the urine can also occur. Treatment is not without potential complications and veterinarians have differing opinions about whether or not mildly affected dogs should be treated at all.

There are two forms of Cushing's syndrome that can be differentiated via blood testing and abdominal imaging (ultrasound, CT scan, or MRI). In approximately 85 percent of Cushingoid dogs, the pituitary gland (in the brain) produces too much ACTH (adrenocorticotropic hormone) that, in turn, stimulates both adrenal glands to enlarge and produce too much cortisone. (Interestingly, levels of adrenaline—also produced within the adrenal gland—remain normal.) This "pituitary-dependent" version is treated with medication (a few different ones are available) to decrease the production of cortisone. In gen-

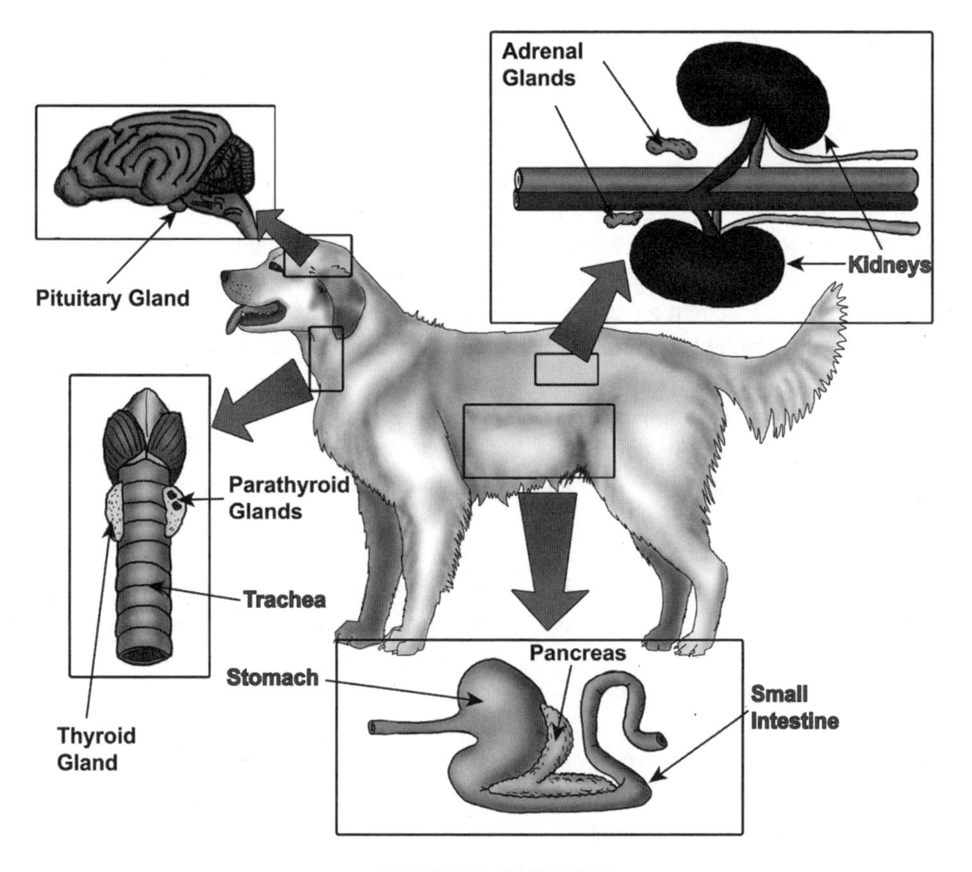

ENDOCRINE SYSTEM

eral, the prognosis for pituitary-dependent disease is good with careful treatment and monitoring, often under the supervision of a veterinary internist.

The second, less common form is described as "adrenal-dependent." Excessive cortisone is produced by a tumor within one of the adrenal glands—approximately half of these tumors are found to be benign while the other half malignant. Unlike the pituitary-dependent form of the disease, which involves symmetrical adrenal enlargement, the adrenal-dependent version has just one enlarged adrenal gland. In this case, the other goes into a "dormant state" in response to the overabundance of cortisone in the bloodstream, often atrophying (shrinking in size) the same way a muscle that is not being used does.

The treatment of choice for adrenal-dependent disease is surgical removal of the affected adrenal gland, although if the tumor is cancerous, successful removal may not be possible because of involvement with other vital structures. If the affected adrenal gland is benign and can be successfully removed, the prognosis is excellent.

Questions for Your Vet

❸ **Is it necessary to treat my dog's Cushing's syndrome?**
❷ **Which form of the disease does he have?**
❸ **If he has a tumor in the adrenal glands, how do we know if it's benign or malignant? Can it be successfully removed surgically? If not, what are other treatment options?**
❹ **If he has pituitary-dependent disease, which medication do you prefer to treat with, and why?**
❺ **Does my dog have high blood pressure or excessive protein leakage into his urine? If so, how should they be managed?**
❻ **How long will it take for symptoms to resolve?**
❼ **If we treat my dog for Cushing's syndrome, is it likely that other medical issues that have been "masked" by the extra cortisone (for example, arthritis or allergies) will surface?**

Diagnosis: Diabetes insipidus

Unlike diabetes mellitus (DM—see p. 301), this version of diabetes has nothing to do with blood sugar levels. *Diabetes insipidus (DI)* is caused by an abnormality of *vasopressin*, a hormone that is stored within the pituitary gland and exerts its effect on the kidneys. Dogs with DI commonly experience a lack of vasopressin production; less often the disease is caused by unresponsive kidneys *(nephrogenic DI)*.

In normal dogs, higher levels of vasopressin are released from the pituitary gland when the body needs to retain water (produce less urine) as during exercise, or when experiencing increased ambient temperature or inadequate water consumption. Dogs with DI cannot conserve water and run the risk of dehydration, especially when not given free access to water. They drink copious amounts in response to their body's production of massive amounts of urine. Increased thirst and urine production are telltale symptoms of this disease.

Diagnosis is confirmed via a study called a *water deprivation test*, or with a *therapeutic trial* (vasopressin is administered to see if the massive thirst and urine output resolve). Brain trauma and brain tumors are possible causes of DI, but often an underlying cause cannot be found. Imaging of the brain (MRI or CT scan) is the test of choice for ruling out a brain tumor. Treatment of DI involves lifelong replacement of vasopressin either given orally in pill form or as a liquid eyedrop. Unless a brain tumor is discovered, with adequate treatment the long-term prognosis is excellent.

Questions for Your Vet

❶ **Should we consider brain imaging (MRI or CT scan) to rule out the possibility of a brain tumor?**
❷ **Which form of vasopressin (pills or eyedrop) do you recommend?**

❸ **How will we know when we are giving the correct dose of vasopressin? Do we need to treat until my dog's thirst level becomes completely normal?**

Diagnosis: Diabetes mellitus (sugar diabetes)

Diabetes mellitus (DM) is caused by a deficiency of the hormone insulin (normally produced in the pancreas) that results in the inability of the body's tissues and organs to properly utilize glucose for energy needs. This causes a dog to "starve in the midst of plenty"—lose weight in spite of a good appetite. Canine diabetes resembles type I diabetes in people in that treatment requires lifelong insulin given by injection (as opposed to the oral medication and change in diet and exercise that can control type II diabetes in people). Causes of diabetes include: genetic susceptibility; pancreatitis (inflammation of the pancreas); destruction of insulin-producing cells by the body's immune system; cortisone-containing drugs; and other hormonal imbalances. Unspayed female dogs and obese dogs are predisposed to diabetes.

Diabetes is diagnosed by a simple blood test that documents a high blood sugar level indicating glucose is not being properly used. Typical diabetic symptoms include: increased appetite; weight loss; and excessive thirst and urine output. Diabetic dogs are also prone to cataract formation, skin and urinary tract infections, and Cushing's syndrome (see p. 298). A serious complication of diabetes called *ketoacidosis* causes decreased appetite, vomiting, profound lethargy and, at its worst, coma. If you see such symptoms, your dog warrants emergency medical attention.

Determining the correct insulin dose to give requires careful monitoring. Too little insulin does not control diabetic symptoms and too much can cause a dangerously low blood sugar concentration *(hypoglycemia)*, symptoms of which can include weakness, trembling, incoordination, and even seizures. Adjustments to the dose are based on how your dog is acting and feeling in conjunction with regular tests that monitor blood sugar levels. This testing can be done in the veterinary clinic, or sometimes at home, the same way human diabetics monitor their blood sugar levels. Your vet might also recommend monitoring your dog's *blood fructosamine level,* which is influenced by how often the blood sugar value was within "normal" range throughout the preceding three weeks. Additionally, at-home monitoring for urine *ketones* (if present in the urine, would signal ketoacidosis) might be recommended.

The prognosis for DM is variable; some dogs are harder to regulate than others. With careful treatment and monitoring some dogs achieve normal life expectancy.

Questions for Your Vet

❶ **Do we know what caused my dog's diabetes?**
❷ **Does my dog have ketones in his urine?**
❸ **What do I need to know about handling and administering insulin properly?**
❹ **What should I do should if symptoms of hypoglycemia occur?**

5. **How should my dog's diet and feeding schedule be altered?**
6. **Should the insulin dosage be altered if he doesn't eat normally?**
7. **How much leeway do I have when it comes to the timing of the insulin injections?**
8. **Is it okay to periodically skip a dose of insulin?**
9. **What should I do if the insulin injection doesn't go well (my dog moves or I accidentally squirt some of the insulin outside the skin)?**
10. **Should his exercise level be altered?**
11. **Should I test his urine for glucose levels and ketones? If so, which findings should prompt me to contact you?**
12. **Is it possible for me to measure blood sugar levels at home?**
13. **Does my dog have cataracts caused by diabetes? If so, should I take him to see a veterinary ophthalmologist?**

❊ **Don't Forget!** Check p. 283 for Universal Questions, and pp. 148 and 152 for questions when dealing with anesthesia and surgery.

Diagnosis: Hyperparathyroidism

Hyperparathyroidism is an overproduction of parathyroid hormone (PTH), which regulates the balance of calcium and phosphorus within the body (see information on *under*production, p. 303). Too much PTH causes an elevated blood calcium level and its associated symptoms include: malaise; weakness; vomiting; diarrhea; increased thirst; loss of appetite; and sometimes kidney failure.

The most common cause of increased PTH production is a benign growth in one of the four parathyroid glands. Ultrasound of the neck often clarifies which gland is the "overachiever." The diagnosis is confirmed by measuring an increased blood PTH level. Surgical removal of the gland or ultrasound-guided injection of ethanol directly into the gland to destroy it *(parathyroid gland ablation)* are the two treatment options. If the calcium value is quite high or the dog is in kidney failure, medical therapy to reduce the blood calcium level is required before surgery or gland ablation.

Dogs commonly experience a temporary low blood calcium level immediately following the removal or destruction of the overactive parathyroid gland. It takes anywhere from a few days to a few weeks for the other three "slumbering" parathyroid glands to wake up and recognize the need for PTH production. Oral or injectable administration of supplementary calcium—with or without vitamin D—is necessary to prevent the symptoms of this transient low blood calcium.

Questions for Your Vet

1. **What are the potential risks and benefits of the treatment options?**
2. **Was my dog's kidney function harmed by the high blood calcium level?**
3. **How will he be monitored post-operatively?**

4. **How frequently will you monitor his calcium level in the short- as well as long-term?**

Diagnosis: Hypoparathyroidism

Hypoparathyroidism is caused by underproduction of parathyroid hormone (PTH), the hormone that is the primary regulator of calcium levels in the body (see information on *over*production, p. 302). Without it, blood calcium levels plummet, resulting in: muscle twitching and tremors; anxiety; vomiting; diarrhea; and seizures. The disease is thought to be due to damage done to the parathyroid tissue by the body's own immune system. Documentation of a low blood PTH level confirms the diagnosis.

Treatment is lifelong and consists of replacement calcium and vitamin D (the latter is necessary for the absorption of calcium from the gastrointestinal tract into the bloodstream). Depending on the severity of the initial symptoms, hospitalization with 24-hour monitoring may be warranted until blood calcium levels stabilize. With careful treatment and monitoring, the prognosis is excellent.

Questions for Your Vet

1. **Is my dog in need of 24-hour care until his calcium level becomes stabilized? If so, who will monitor him through the night?**
2. **How will I know if I am giving too much or too little calcium and vitamin D?**
3. **How frequently should his calcium be monitored on an ongoing basis?**

Diagnosis: Hypothyroidism

An underproduction of thyroid hormone by the thyroid glands is called *hypothyroidism*. Thyroid hormone controls the body's metabolic rate. Too much thyroid hormone "revs it up," and too little—as in this case—slows it down. Hypothyroidism causes: sluggish behavior; unexplained weight gain; "heat-seeking" behavior (the dog spends more time in the sun, or by the radiator or woodstove); breeding problems; and changes in skin and hair coat. Diagnosis is confirmed by blood tests that measure thyroid hormone levels.

Treatment consists of daily, lifelong supplementation with oral thyroid replacement hormone. Follow-up blood tests are performed on a regular basis to be sure that the dosage is neither too high nor too low. The good news is that treatment is quite inexpensive and the prognosis usually excellent.

Note: A number of different medications and nonthyroidal diseases can create blood test results suggestive of hypothyroidism even when thyroid function is completely normal. This is referred to as "sick euthyroid syndrome." The take-home message is that your dog should not be treated for hypothyroidism based solely on blood test results—he must exhibit related symptoms.

Questions for Your Vet

1. **How do we know that my dog is truly hypothyroid and not "sick euthyroid"?**
2. **Is it okay to use the generic form of thyroid medication?**
3. **How will I know if my dog is getting too much or too little thyroid hormone?**

4. **How long will it take for the hypothyroid symptoms to resolve?**
5. **How long after I give my dog his morning thyroid supplement should the blood sample be taken to retest his thyroid level?**

Gastrointestinal Diseases

Introduction to the Digestive System

When a dog eats, the food moves through and is processed within the gastrointestinal (GI) tract. First it is passed via the esophagus (located within the chest cavity) into the stomach where the food is "pulverized" into a gruel-type consistency. After leaving the stomach, the material negotiates several feet of small intestine. This is where nutrients from the food are absorbed into the bloodstream. Finally, the food material is transported through the colon (large intestine) where water is reabsorbed. The colon terminates at the anus.

QUICK REFERENCE

Which Specialist Is Right for My Dog?

If referral to a specialist is needed, a board certified internist or surgeon will be chosen based on whether the dog is in need of medical or surgical diagnostics and treatment.

Diagnosis: Acute gastroenteritis (acute gastritis, dietary indiscretion, garbage can enteritis)

Sooner or later, just about every dog experiences a bout of *acute gastroenteritis*, inflammation of the gastrointestinal (GI) tract. Causes include: eating or drinking something unintended for canine consumption (garbage, plants, medications, compost, household cleaning products); food intolerance; toxins; and infections. In some cases, the underlying cause is never determined. The most common symptoms include: vomiting; diarrhea; lethargy; and decreased appetite. Diagnosis is made based on known ingestion of something "out of the ordinary" and by ruling out other diseases.

Treatment of acute gastroenteritis includes withholding food for 24 hours to give the GI tract a chance to recover, and medication to reduce nausea. Supplemental fluids can be administered when dehydration is detected. Once vomiting has resolved, small frequent meals of a bland, easily digestible diet (plain, boiled, white rice; lean, cooked, white-meat chicken or turkey; or a prepared diet prescribed by your vet) are offered. A rapid recovery time (typically within 24 to 48 hours) is anticipated.

Questions for Your Vet

1. **Do we know what caused this?**
2. **How severely is my dog affected? Is he better off being treated at home or in the hospital?**
3. **How long do you anticipate it will take for the vomiting and diarrhea to fully resolve?**
4. **When can I resume feeding my dog his normal diet?**

Diagnosis: Anal gland disease (anal sac impaction, anal sac infection)

Just like the human appendix, a dog's two anal sacs are only good for causing problems! Located at the ten o'clock and two o'clock positions, just underneath the skin surface of the anus, the anal glands produce a horrific smelling pasty material that accumulates within the sacs surrounding them. This material is emptied from the sacs via two small ducts that lead to the surface of the anus. Evacuation of the sacs usually occurs in conjunction with the pressure of a bowel movement. Some dogs experience an impaction (the material cannot be eliminated) or a bacterial infection within one or both sacs. Symptoms include: licking at the anal region; "scooting" behavior; swelling around the anus; reluctance to sit; and awful smelling secretions from the anus that may contain blood.

The cause of anal sac disease is uncertain, but obesity, various skin conditions (such as seborrhea), and food allergies may be predisposing factors. Treatment includes manual expression of the anal sac, and if infection is present, antibiotics. For dogs that are "repeat offenders" frequent manual anal sac emptying or surgical removal of the affected sac are reasonable options.

Questions for Your Vet

1. **Are one or both anal sacs affected?**
2. **Is infection present?**
3. **Should I ask the groomer to empty my dog's anal sacs on a regular basis?**
4. **What can I do at home to prevent this from recurring? Can you teach me how to empty my dog's anal sacs?**

Diagnosis: Gastric dilatation/volvulus (gastric torsion, bloat)

In dogs with *gastric dilatation/volvulus (GDV)* the stomach distends with gas (bloat), then twists on itself (volvulus), pinching off its connection with the esophagus and small intestine, and compromising normal blood flow to the stomach. GDV is relatively common in large and giant deep-chested dog breeds. The cause of GDV is unknown; diet and the timing of feeding and exercise may play a role. Symptoms include an abrupt onset of nonproductive retching, abdominal distention, and weakness. The diagnosis is confirmed with X-rays of the abdomen.

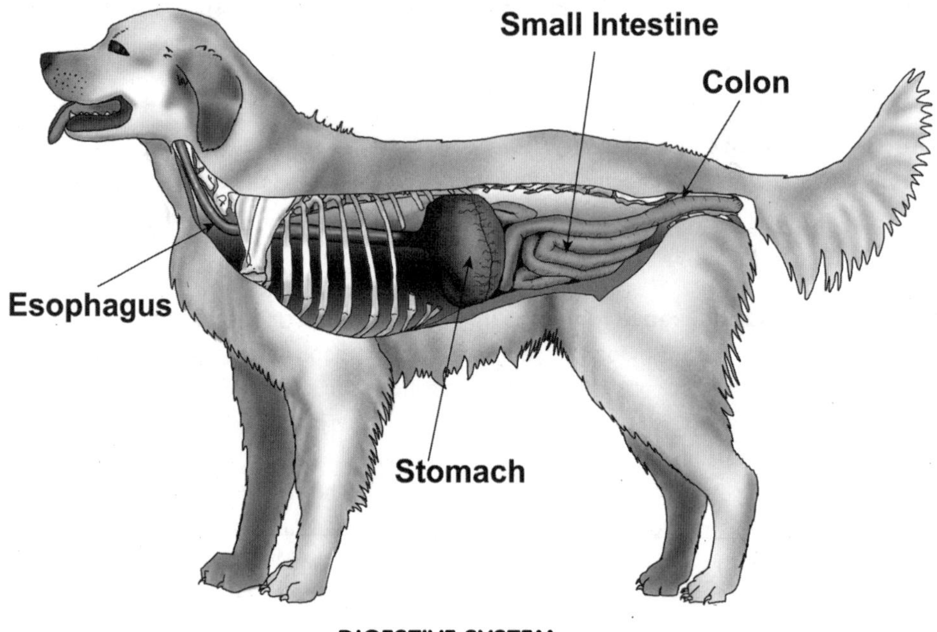

DIGESTIVE SYSTEM

GDV should always be treated as an emergency, as it requires stabilization with intravenous fluids, removal of gas from the stomach, and surgery to de-rotate the stomach. During surgery, a *gastropexy* (a portion of the stomach wall is permanently sutured to the body wall) is performed to prevent recurrence of the volvulus (torsion). The surgeon may need to remove a portion of the stomach and/or spleen because of compromised blood flow. Common post-surgical complications include: heart rhythm abnormalities; aspiration pneumonia; stomach ulcers; blood-clotting problems; and recurrent stomach bloating with gas. The prognosis correlates with how quickly surgery can be performed, the severity of post-operative complications, and whether or not it was necessary to remove a portion of the stomach.

Questions for Your Vet

1. **How compromised is my dog going into surgery? What are his chances for a successful outcome?**
2. **Was it necessary to remove a portion of the stomach? If so, how much?**
3. **Is my dog experiencing any post-operative complications? How are they being treated?**
4. **How is pain being managed?**

5. **How long will he likely need to stay in the hospital?**
6. **What can I do to prevent future stomach bloating?**

❊ **Don't Forget!** Check p. 283 for Universal Questions, and pp. 148 and 152 for questions when dealing with anesthesia and surgery.

Diagnosis: Inflammatory bowel disease (colitis)

Inflammatory bowel disease (IBD) is a group of disorders characterized by an abnormal accumulation of inflammatory cells (various types of white blood cells) within the walls of the gastrointestinal (GI) tract that disrupt normal gut function. The actual cause of IBD is uncertain but is likely multifactorial. Possible causes include: GI infections or parasites; allergies to food or normal gut bacteria; stress; drug reactions; immune system abnormalities; and genetics.

Any portion of the GI tract can be affected. Symptoms depend on the location, type and severity of the inflammation and include: vomiting; diarrhea; blood or mucous in the stool; straining to have a bowel movement; flatulence; decreased appetite; lethargy; and weight loss. When severe or chronic, IBD can cause: malnutrition; weakness; low blood protein levels; overgrowth of bacteria within the small intestine; and changes in other organ systems. The diagnosis is based on gut biopsies and ruling out other GI diseases (infections, parasites, food intolerance, cancer). Biopsies can be collected via surgery or with endoscopy, where a flexible telescopic device is passed down the esophagus into the stomach and small intestine or—coming in from the other direction—through the rectum into the colon.

The actual classification of IBD and the treatment plan are formulated based on the biopsy results. Changes in the diet might be indicated to: reduce the potential for an allergic response; alter fiber content; reduce fat content; and enhance digestibility. Several medications are used, alone or in combination, to reduce inflammation, inhibit the immune system, and restore normal intestinal bacterial populations. Just as in people with IBD, the disease is typically chronic. Symptoms may wax and wane, but long-term—if not lifelong—therapy is usually necessary.

Questions for Your Vet

1. **What part of my dog's gastrointestinal tract is affected by the IBD?**
2. **How severe is the disease based on biopsy results?**
3. **Do the biopsies give us an idea of the underlying cause?**
4. **Does my dog have any problems secondary to the IBD?**
5. **Can he have his usual treats and chew toys?**
6. **Should I change his diet? Should I alter the frequency of his meals?**
7. **How long will it take to know whether or not the treatment is working?**

Diagnosis: Ingested foreign body

Dogs delight in eating stuff they shouldn't and sometimes these items become lodged within the gastrointestinal (GI) tract, where they can create local irritation, cause an obstruction, or perforate the bowel wall. Symptoms produced are dependent on the location of the foreign body—higher up in the GI tract is more likely to cause regurgitation or vomiting (see explanation of these two symptoms on p. 280); lower down is more likely to cause diarrhea in conjunction with vomiting. Perforation results in lethargy, fever, and pain. The diagnosis of an esophageal or gastrointestinal foreign body relies on imaging studies (X-rays, ultrasound, CT scan, MRI). When imaging does not help with diagnosis, *endoscopy* (passage of a telescope device into the GI tract) or exploratory surgery is needed.

Some foreign bodies are small enough to pass through on their own and follow-up X-rays are used to document their progress. If the object is in the stomach—depending on what the object is—your vet may recommend induction of vomiting in the hopes of bringing it back up. Surgery is the most common treatment, but removal by endoscopy is often successful when the foreign body is in the esophagus, stomach, or upper small intestine and is a size and shape that can readily be grasped with endoscopic snares. Endoscopy is the preferred method for removing anything from the esophagus because surgical scarring is a common and serious complication at this site. Prognosis depends on the patient's condition at the time of treatment and the degree of damage sustained by the GI tract.

Questions for Your Vet

1. **Where is the foreign body located? What do you think it is?**
2. **Is it causing an obstruction?**
3. **Does it need to be removed?**
4. **Do you think it has caused a perforation?**
5. **Is removal via endoscopy an option?**

Diagnosis: Intestinal parasites (giardia, roundworms, whipworms, hookworms, coccidia)

Intestinal parasites are commonly diagnosed, particularly in puppies. The prevalence of parasitic infection varies with geographic location and climate. They are contagious from dog to dog via the feces, and some are transmissible between species, including dog to human. Others are transmitted from mother to puppies during pregnancy and nursing. Stress (experienced during weaning, performance, and showing) and concurrent diseases increase the likelihood of infection by suppressing normal immune defenses.

Gastrointestinal parasites universally cause diarrhea. Vomiting and loss of appetite occur less frequently. Chronic infection can cause weight loss and an overall unthrifty appearance. Adult worms live in the intestinal tract and are not usually present in the stool—but their eggs (ova) are. For all parasites other than giardia, the eggs are the basis for the diagnosis when viewed under the microscope. A giardia antigen test (also performed on a stool sample) is used when giardia is suspected.

Treatment consists of oral medication to kill the parasites and prevent reinfection. The prognosis is excellent.

Questions for Your Vet

1. **What type of parasite does my dog have?**
2. **Does he have an underlying intestinal disease that may have predisposed him to a parasite problem?**
3. **What should I do to prevent reinfection?**
4. **Should all of my dogs and cats be treated?**
5. **When should a stool sample next be checked to be sure that the parasites are gone?**
6. **Should a stool specimen be checked for parasites on a routine basis?**
7. **Am I or other family members at risk for getting the infection? If so, what preventive measures should I take?**

Diagnosis: Megaesophagus

The esophagus is the muscular tube that moves food and water from the mouth down to the stomach by coordinated contractions called *peristalsis*. When *megaesophagus* occurs, the esophagus dilates and the muscles become flaccid and weak. Most cases of megaesophagus are idiopathic (a cause cannot be found). Known causes of megaesophagus include: neurological disorders; *myasthenia gravis* (see p. 331); *esophagitis* (inflammation of the esophagus); lead toxicity; and some hormonal imbalances.

Megaesophagus results in regurgitation. Unlike vomiting, when regurgitation occurs, undigested food, water, and saliva that hasn't been successfully transported to the stomach suddenly spews forth from the mouth without any retching or other warning. Because of the lack of warning, aspiration of some of the regurgitated material into the lungs (aspiration pneumonia) is the most common and life-threatening complication of megaesophagus. Symptoms commonly associated with aspiration pneumonia include lethargy, fever, and cough.

The diagnosis of megaesophagus and/or aspiration pneumonia can usually be made with plain X-rays. Other diagnostic tests can be used to rule out known causes of megaesophagus. If an underlying cause cannot be found, treatment consists of altering diet and keeping the dog in an elevated position after feeding to let gravity help transport food into the stomach. The prognosis is variable and largely depends on whether or not aspiration pneumonia occurs.

Questions for Your Vet

1. **Do we know what caused the megaesophagus? If so, what can we do to treat it? If not, what can we do to determine the cause?**
2. **Does my dog currently have evidence of aspiration pneumonia? If so, how should it be treated?**

3. **How should I alter his diet and feeding schedule?**
4. **Should his activity level be changed?**
5. **Should X-rays be repeated in the future to see if the megaesophagus and pneumonia (if present) have resolved?**
6. **Which symptoms should I be watching for that may be indicative of aspiration pneumonia? If they occur, should my dog be seen on an emergency basis?**

Diagnosis: Small intestinal bacterial overgrowth (bacterial overgrowth, antibiotic responsive diarrhea)

Many species of bacteria are normal inhabitants of the canine small intestine. They are a vital component of normal food processing and digestion. *Small intestinal bacterial overgrowth (SIBO)* occurs when the bacteria undergo a population explosion resulting in a variety of gastrointestinal symptoms, including: diminished appetite; gurgly gut sounds (*borborygmus*); diarrhea; and weight loss. SIBO most commonly occurs secondary to another gastrointestinal disease (inflammatory bowel disease, cancer, parasites, exocrine pancreatic insufficiency). Occasionally, SIBO occurs as a primary disease process, particularly in German Shepherds.

Breath hydrogen testing and quantification of bacteria within the small intestine can be used to confirm diagnosis of SIBO, but they are rarely performed because of their level of technical difficulty. More commonly, diagnosis is based on clinical signs and abnormal blood levels of *folate* (elevated) and *cobalamin* (decreased). Both folate and cobalamin are forms of Vitamin B.

SIBO is treated with antibiotics (often long-term) and cobalamin supplementation (if levels are significantly decreased) in conjunction with treatment of any underlying bowel disorder. Probiotics, such as *lactobacillus*, may be of benefit to restore normal intestinal bacterial populations. SIBO is not a transmissible disease.

Questions for Your Vet

1. **Do we know what the cause of my dog's SIBO is? If not, what can be done to find out?**
2. **If my dog has a primary gastrointestinal disease that is causing the SIBO, how should it be treated?**
3. **Should I change his diet?**
4. **Will folate and cobalamin testing be repeated? If so, when?**

Hematological Diseases

Introduction to Blood Cells

Hematological diseases affect blood cells. *Red blood cells* transport oxygen around the body and *platelets* are cells responsible for normal blood clotting. *White blood cells* (lymphocytes, neutrophils, and eosinophils) and *macrophages* are involved in the body's normal immune system function. They tend to be "overachievers" in cases of allergic and immune mediated diseases.

> **QUICK REFERENCE**
>
> **Which Specialist Is Right for My Dog**
>
> The diseases listed in this section are all capable of creating some diagnostic and therapeutic challenges. If help or a second opinion is needed, your veterinarian will refer you and your dog to an internal medicine specialist.

Diagnosis: Anemia

A lower than normal red blood cell count is referred to as *anemia*. Red blood cells are responsible for carrying oxygen around the body, so it makes sense that dogs with anemia show decreased stamina, lethargy, and decreased appetite. Symptoms correlate with the severity of the anemia and how slowly or rapidly it developed. Potential causes include: active bleeding; decreased production of new red blood cells within the bone marrow; and destruction (*hemolysis*) of circulating red blood cells.

Anemia is diagnosed via a simple blood test, but more involved testing is needed to determine the cause, treatment, and prognosis. If the anemia is severe, a blood transfusion may be recommended.

Questions for Your Vet

1. **What is the cause of my dog's anemia? If unknown, what testing can be done to determine the cause?**
2. **How severe is the anemia? Is it life-threatening?**
3. **Is it a responsive anemia (i.e., is the bone marrow making new red blood cells)?**
4. **Does my dog need a blood transfusion?**
5. **Should I decrease his activity level?**
6. **Should I alter his diet or provide an iron supplement?**

Diagnosis: Hemophilia (von Willebrand disease, inherited coagulation factor deficiency)

A dog with *hemophilia* has abnormal blood clotting. Normal blood clotting depends on the interaction of multiple blood proteins (*clotting factors*). Dogs with hemophilia are born with an abnormality of one or more of these clotting factors. Hemophilia is suspected when a dog has spontaneous bleeding (no known trauma) or bleeds excessively during surgery. Highly specialized blood testing is used to determine which clotting factor is the culprit. Hemophilia cannot be cured, but active bleeding can usually be controlled with a transfusion rich in blood clotting factors.

Von Willebrand disease (VWD) is the most common form of canine hemophilia and Doberman Pinchers are the number-one breed represented. Not all dogs that test positive for VWD have active bleeding. The prognosis is determined by the severity of the von Willebrand factor deficiency and whether or not the factor present is functional.

Questions for Your Vet

1. **What type of hemophilia does my dog have?**
2. **How severe is the hemophilia?**
3. **What should I do if he has a bleeding episode?**
4. **What should we do if surgery is needed?**
5. **Are there drugs to be avoided that might increase my dog's bleeding tendencies?**
6. **Are there treatments that can reduce the risk of bleeding?**
7. **Will he get better or worse over time?**
8. **Should my dog be spayed/neutered (if not already)?**

* **Don't Forget!** Check p. 283 for Universal Questions, and pp. 148 and 152 for questions when dealing with anesthesia and surgery.

Diagnosis: Immune mediated hemolytic anemia (autoimmune hemolytic anemia)

The immune system protects the body by recognizing and eliminating foreign substances (viruses, bacteria, splinters) while leaving the body's own tissues unharmed. *Immune mediated hemolytic anemia (IMHA)* occurs when the immune system is somehow triggered into believing that its own red blood cells don't belong and should be destroyed (*hemolyzed*). Although new red blood cells are constantly being produced in the bone marrow, the rate of production cannot keep up with the rate of destruction and a severe anemia typically results. Testing seeks to identify what triggered the immune system to behave badly (infection, vaccination, drug, cancer). Even with a thorough search, an underlying cause often cannot be identified.

IMHA is diagnosed on the basis of anemia that is regenerative (lots of baby red blood cells are being released from the bone marrow) along with characteristic changes in the appearance of the red blood cells that indicate they are being damaged by the immune system. Some dogs with IMHA become jaundiced because of the circulating red blood cell breakdown products. A predisposition to formation of blood clots (especially within the lungs) and *immune mediated thrombocytopenia* (see p. 314) commonly accompany IMHA.

Treatment consists of eliminating any underlying cause and giving immunosuppressive medications in hopes of convincing the immune system to leave the red cells alone. Blood transfusions may also be necessary. With most immune mediated diseases, treatment is usually long-term, even lifelong. Unfortunately, the prognosis for IMHA is guarded. Some dogs fail to respond to therapy.

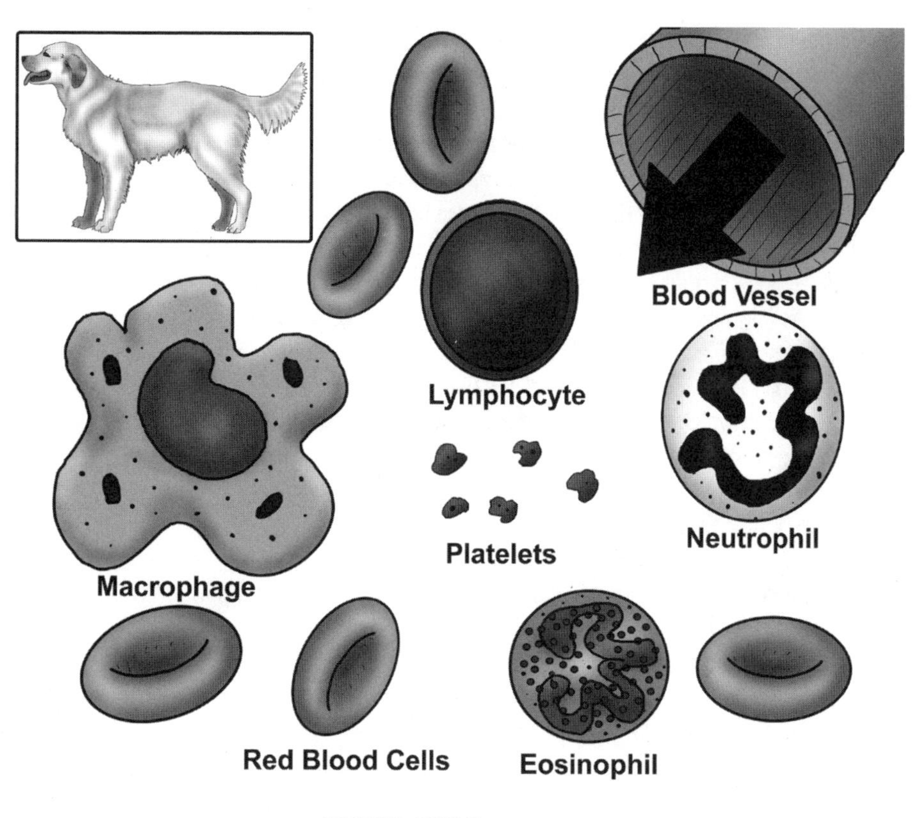

BLOOD CELLS

Questions to Ask Your Vet

1. Do we know what triggered the IMHA?
2. Does my dog have evidence of any other immune mediated diseases?
3. Does my dog need a blood transfusion?
4. Is my dog at risk for blood clot formation? If so, how should this be treated?
5. Should we continue to monitor his red blood cell count even when he is showing no symptoms? If so, how often?
6. Should I keep my dog away from other dogs while he is receiving immunosuppressive therapy?
7. Should his activity level be altered?
8. Should he continue to receive vaccinations?

Diagnosis: Immune mediated thrombocytopenia

Thrombocytopenia refers to a decrease in the number of blood platelets (thrombocytes). Platelets play an important role in blood clotting. When they are in short supply, spontaneous bruising and bleeding occur, sometimes to the extent of causing anemia. Just as the immune system can be triggered to attack red blood cells (see p. 312) it can also turn on the platelets, resulting in *immune mediated thrombocytopenia (IMT)*. This disease is readily diagnosed by measuring the blood platelet count in conjunction with ruling out other causes of thrombocytopenia. IMT commonly occurs in conjunction with IMHA or immune mediated activity directed against the kidneys *(glomerular disease)* or joints *(polyarthritis)*.

Potential triggers of IMT include infectious diseases, vaccinations, medication, and cancer. Treatment of IMT is directed at removing the underlying cause although in many cases, it cannot be determined. When the cause cannot be identified and treated, immunosuppressive medications are used to stop the immune system destruction of the platelets. Blood transfusions are necessary when spontaneous bleeding is severe. The prognosis is variable and treatment is typically long-term.

Questions to Ask Your Vet

1. **Do we know what triggered the IMT?**
2. **Does my dog have evidence of any other immune mediated diseases?**
3. **How severe is his disease?**
4. **Does he need a blood transfusion?**
5. **Should we continue to monitor his platelet count even if he is showing no symptoms? If so, how often?**
6. **Should I keep him away from other dogs while he is receiving immunosuppressive therapy?**
7. **Should his activity level be altered?**
8. **Should he continue to receive vaccinations?**

Hepatic (Liver), Biliary (Gall Bladder), and Pancreatic Diseases

An Introduction to the Hepatic, Biliary, and Pancreatic Systems

The liver is also known as the hepatic system. It serves as a "garbage disposal" in that it rids the body of toxic substances, especially those that are ingested (and dogs certainly eat many unsavory things). The liver is also responsible for manufacturing protein, cholesterol, glucose, and bile for use in the body. The gall bladder stores bile, a detergent-like substance that is released into the small intestine to help with fat digestion. The pancreas is the body's source for insulin and digestive enzymes.

QUICK REFERENCE

Which Specialist Is Right for My Dog?

If a specialist's help is needed with your dog's liver, biliary, or pancreatic disease, your veterinarian will likely want to enlist the help of a board certified internist. For those patients in need of surgery involving the liver, gall bladder, or pancreas, a board certified surgeon should be consulted.

Diagnosis: Acute hepatic injury (liver injury, acute hepatic failure, acute hepatic necrosis)

The liver is the major processing plant for toxins in the body. More commonly than other organs, it sustains significant injury from poisonous substances (mushrooms, chemical compounds); medication (prescription, over-the-counter, anesthetic agents); infectious organisms (bacteria, viruses, fungi); and diseases affecting other organs (inflammatory bowel disease, pancreatitis, hemolytic anemia). Liver damage can also occur with physical trauma and heat stroke. Although the liver has a great deal of reserve and is capable of regeneration, the damage may be severe enough to cause a state of liver failure.

Symptoms of acute liver injury include lethargy, vomiting, diarrhea, and decreased appetite, often with a sudden onset. A presumptive diagnosis can be made based on observed ingestion of something known to be toxic to the liver. A liver biopsy provides a definitive diagnosis. Liver injury results in increased liver enzyme values in the blood, but these are nonspecific, and consistent with any form of liver disease. However, these liver enzymes are useful for monitoring therapy. Treatment involves supportive therapy (typically in a hospital setting) to allow the liver to repair and regenerate, as well as prevention or control of any complications associated with liver failure. Also, any underlying disease is treated when possible. The prognosis varies with the cause of the liver injury and the degree of liver damage.

Questions to Ask Your Vet

1. **Do we know what caused injury to the liver? If so, is there a specific treatment or antidote to stop ongoing liver damage?**
2. **Is my dog in liver failure?**
3. **How long a hospital stay do you anticipate?**
4. **Who will monitor my dog overnight?**
5. **When will more blood tests be done to help us know if things are improving or worsening?**
6. **Is the damage likely to be permanent?**
7. **If toxicity is suspected, should my other dogs be tested for liver damage?**

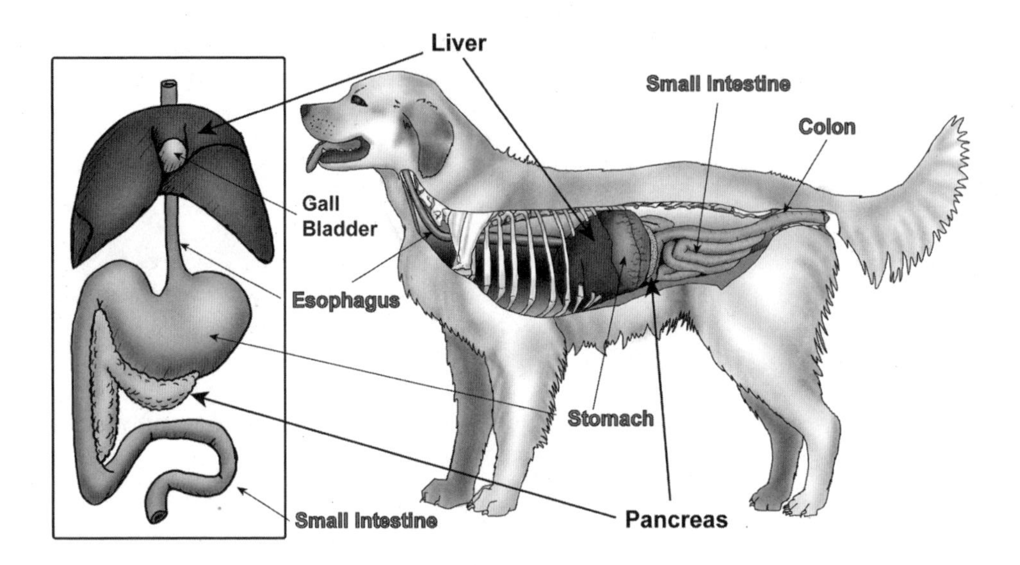

HEPATIC, BILIARY, AND PANCREATIC SYSTEM

Diagnosis: Exocrine pancreatic insufficiency

The pancreas has a number of functions, one of which is to secrete digestive enzymes into the small intestine. *Exocrine pancreatic insufficiency (EPI)* is the failure of the pancreas to produce these enzymes. EPI can occur as an aftermath of pancreatitis (inflammation of the pancreas), but in most cases, the cause of this disease is unknown. EPI occurs most commonly in German Shepherds.

Dogs with EPI appear to be starving—they have a voracious appetite, but continue to lose weight. They have diarrhea, and massive stool quantity—due to the absence of digestive processes, everything that goes in one end simply comes out the other.

The *trypsin-like immunoreactivity (TLI) test* assesses the presence of digestive enzymes within the bloodstream. A markedly reduced TLI result is consistent with the diagnosis of EPI. Many dogs with EPI also have *cobalamin* (vitamin B-12) deficiency and/or *small intestinal bacterial overgrowth (SIBO)*. These secondary problems must be identified and treated because they can continue to cause diarrhea even after the EPI has been successfully managed. Treatment of EPI is lifelong and consists of mixing the dog's meal with a pancreatic enzyme to "predigest" the food.

Questions for Your Vet

1. **Should I change my dog's diet?**
2. **How much should I feed him?**
3. **How much weight does he need to regain?**

❹ **Does he have small intestinal bacterial overgrowth or cobalamin deficiency? If so, how should they be treated?**

❊ **Don't Forget!** Check p. 283 for Universal Questions, and pp. 148 and 152 for questions when dealing with anesthesia and surgery.

Diagnosis: Gall bladder disease (cholecystitis, biliary mucocele, gall bladder stones)

The gall bladder is a reservoir for bile, a substance made in the liver and excreted into the small intestine to aid with digestion. Although common in people, *gall bladder stones* are rare in dogs, usually cause no symptoms, and often do not require therapy. *Cholecystitis* (inflammation of the gall bladder) is the most common *gall bladder disease* in dogs. When associated with liver inflammation it is referred to as *cholangiohepatitis*. Cholecystitis may or may not be associated with a bacterial infection. A *biliary mucocele* is likely a variation of cholecystitis in which the gall bladder fills with a thick mucoid material that often leads to obstruction of the bile duct. The biggest concern with gall bladder disease in dogs is the potential for rupture and release of bile into the abdominal cavity. The resulting *peritonitis* (inflammation of the abdominal cavity) can be life-threatening.

Symptoms of gall bladder disease include: vomiting; diarrhea; lethargy; decreased appetite; weight loss; and jaundice (yellow hue of the skin and gums). Abdominal ultrasound is the most reliable noninvasive means of diagnosing gall bladder disease. The decision whether to treat the dog medically versus surgically (removal of the gall bladder) depends on the severity of symptoms and ultrasound appearance of the gall bladder. Antibiotics and medication to improve bile flow are used when medical therapy is appropriate.

Questions for Your Vet

❶ **What type of gall bladder disease does my dog have?**
❷ **Is there a need for treatment at this time? If so, is he a better candidate for medical or surgical therapy?**
❸ **If we opt for medical therapy, how will we know whether or not he is improving?**
❹ **Does he have liver disease in addition to gall bladder disease? If so, what kind and how is it best treated?**
❺ **Should his diet be changed?**

Diagnosis: Hepatitis (inflammatory liver disease, infectious hepatitis, chronic active hepatitis)

Hepatitis is inflammation of the liver. There are several different causes of hepatitis (infectious and noninfectious), but all produce similar symptoms, including: vomiting; diarrhea; decreased appetite; weight loss; and lethargy. The diagnosis is made by blood tests and liver biopsy. Although hepatitis causes increases in liver enzyme values in the bloodstream, this is a nonspecific change common to many types of liver disease. The enzyme values *are* useful when monitoring treatment, which includes medication to: reduce inflammation; counteract the symptoms of liver failure; and support liver health. In spite of appropriate treatment, hepatitis can be progressive, ultimately leading to irreversible liver failure (see p. 318).

The most common type of inflammatory liver disease is *chronic active hepatitis*. The underlying cause isn't certain, but the inflammation is thought to be a result of an overactive, misdirected immune system. As the name implies, this is a chronic condition requiring long-term—if not lifelong—management.

A number of infectious agents can cause hepatitis. The list includes: bacteria; fungi; rickettsiae (tick-borne diseases); and viruses (although *not* the same viruses that cause hepatitis in people). Medication is given to treat any associated infection, when recognized.

Some dogs are predisposed to copper accumulation within the liver resulting in chronic hepatitis and irreparable liver damage. This is a result of an inherited metabolic defect (Doberman Pinchers, Labrador Retrievers, and Bedlington Terriers are the breeds most commonly affected). Drugs that *chelate* (bind) copper are given in conjunction with the other types of therapy mentioned above.

Questions to Ask Your Vet

1. **Do we know what caused the hepatitis? If so, how is it best treated?**
2. **How severe is the hepatitis? What is the likelihood of it leading to irreversible liver failure?**
3. **Which is better for my dog, treatment at home or treatment in the hospital?**
4. **When will liver enzyme values next be tested?**
5. **Should I revise my dog's diet and feeding schedule? If so, how?**
6. **Should his activity level be restricted?**
7. **Should I discontinue any of the routine medications he receives for flea and heartworm control?**

Diagnosis: Liver failure (cirrhosis)

The liver is a vital organ that performs many functions. It eliminates toxins absorbed by the gastrointestinal tract; participates in digestion; stores energy for the body; and manufactures sugar, fat, and proteins (including those needed for normal blood clotting). Liver injury, toxicity (some household chemicals, poisonous plants and medications—see p. 374), infection, or inflammation—when severe or chronic enough—can cause liver failure. The failure can be transient (the liver repairs itself) or permanent, in which case normal liver tissue is replaced by scar tissue (cirrhosis).

Symptoms of liver failure include: vomiting; diarrhea; decreased appetite; lethargy; weight loss; and increased thirst and urine output. Fluid accumulation in the abdomen, spontaneous bruising and bleeding, and neurological symptoms (*hepatic encephalopathy*) may also occur. A liver biopsy helps determine whether or not cirrhosis is present and, in some cases, the underlying cause of the failure. In addition to treating the cause of the liver failure, therapy includes: a low protein diet, vitamin supplements, antioxidants; medication to prevent stomach ulcers; vitamin K to prevent bleeding; and drugs that inhibit ongoing scar tissue formation in the liver. Noncirrhotic liver failure is potentially reversible, depending on the prognosis for the underlying disease process. Cirrhosis is irreversible and carries a poor prognosis.

Questions for Your Vet

1. **Do we know what caused my dog's liver failure?**
2. **Is the liver failure potentially reversible? Does he have cirrhosis?**
3. **If my dog has cirrhosis, which symptoms might improve with therapy?**
4. **How should his diet be changed?**
5. **Should I alter his activity level?**
6. **Should I discontinue any of the routine medications my dog receives for flea and heartworm control?**
7. **When will blood work next be rechecked?**
8. **What symptoms should prompt an emergency visit?**

Diagnosis: Pancreatitis

Among its functions, the pancreas produces digestive enzymes that are released into the small intestine. Sometimes these same enzymes are released into the pancreatic tissue causing inflammation we call *pancreatitis*. Recent ingestion of a fatty meal is commonly reported and some medications have the potential to trigger pancreatitis, but often the cause is not evident. Symptoms include: decreased appetite and activity; abdominal pain; and vomiting. Diabetes, kidney failure, heart rhythm disturbances, obstruction of the bile duct, and blood clotting abnormalities are potential life-threatening complications of pancreatitis.

Pancreatitis is diagnosed with a combination of blood tests (which can cause false positive results) and abdominal ultrasound. Dogs with pancreatitis are best treated with intravenous fluids and, initially, receive nothing by mouth so as to control vomiting and avoid stimulating pancreatic enzyme secretion. When started back on food, a fat-free diet is preferred. Treatment also includes medication to control pain and nausea. With appropriate management, most dogs with pancreatitis fully recover. Some dogs are predisposed to recurrent pancreatitis. In such cases, maintenance of a low-fat diet is the best means of prevention.

Questions for Your Vet

1. **How do we know my dog has pancreatitis? How reliable is the evidence we have?**
2. **Is he having any serious complications from the pancreatitis?**
3. **How long will food and water likely be withheld?**
4. **How long do you think he will need to stay in the hospital? Who will monitor him overnight?**
5. **Do you think my dog is experiencing pain?**
6. **What diet do you recommend when he comes home? What diet do you suggest long-term?**
7. **What can be done to prevent future bouts of pancreatitis?**

Diagnosis: Portosystemic shunt (liver shunt, microvascular dysplasia)

A *portosystemic shunt (PSS)* is a defect whereby blood is shunted *around* rather than *through* the liver, which effectively eliminates the liver's normal function in maintaining health. This results in a build-up of metabolic waste materials and a failure to produce important substances (blood clotting factors, albumin, glucose, urea, cholesterol) normally manufactured by the liver.

Shunts may occur as congenital defects or may be acquired secondary to chronic liver disease. The abnormal vessel(s) can be outside or within the liver itself. Dogs with PSS may have: neurological symptoms (abnormal behavior, weakness, circling, pacing, seizures, coma); vomiting; diarrhea; increased thirst; and decreased appetite. Protein in the diet tends to exacerbate these symptoms, which are often worse after mealtime. Bladder stones are a common finding in dogs with PSS. Additionally, affected dogs tend to have prolonged anesthetic recovery times—decreased blood flow to the liver impairs this organ's normal ability to metabolize and eliminate anesthetic drugs from the body.

PSS is suspected based on symptoms, decreased liver size and characteristic blood test changes. Confirmation is obtained via ultrasound, X-rays following an injection of contrast material, or a nuclear medicine study. A liver biopsy is an important part of the diagnostic workup; rerouting blood flow through the liver might not be beneficial if the liver is diseased.

Surgery is the treatment of choice for young dogs with a single (rather than multiple) shunting vessel. When surgery is successful, normal blood flow through the liver is restored. Bladder stones, if present, can be removed at the same time. Sometimes, surgery is not recommended because there are multiple shunts, the liver is cirrhotic (nothing but scar tissue), or the dog is not a good candidate for surgery. In these cases, medical therapy is advised and consists of: a low protein diet; antibiotics; nutritional supplements; and medication to reduce the level of toxins normally cleared by the liver. The prognosis for surgically corrected shunts is variable. Medical therapy can provide a period of good quality time, but the long-term prognosis is poor.

Microvascular dysplasia (MVD) is a relatively common congenital defect that occurs primarily in small terrier breeds. MVD is caused by shunting of blood on a microscopic level throughout the liver. Dogs with this disease often have blood test results similar to dogs with PSS, but many dogs remain free of symptoms. The treatment and prognosis are based on the presence and severity of symptoms. MVD is always managed medically rather than surgically.

Questions for Your Vet

1. **Is the shunt a type that was present at birth?**
2. **Does my dog have a portosystemic shunt (PSS) or microvascular dysplasia (MVD)?**
3. **What does the liver biopsy show?**
4. **Is my dog a good candidate for surgical correction of the shunt? If not, why?**
5. **Does he have urinary bladder stones? If so, can they be surgically removed?**
6. **Is it likely that his symptoms would improve with medical rather than surgical therapy? If so, for how long?**

Infectious Diseases

Introduction to Infectious Diseases

The most common infectious diseases in dogs are caused by bacteria, viruses, and fungal organisms. Rickettsiae are a type of bacteria transmitted to dogs via ticks.

> **QUICK REFERENCE**
> **Which Specialist Is Right for My Dog?**
>
> The type of specialist recruited to help with the diagnosis and treatment of an infectious disease depends on which body system(s) is most affected. Your veterinarian may recommend referral to an internist, ophthalmologist, and/or neurologist.

Diagnosis: Distemper (canine distemper)

Distemper is a highly contagious, viral disease that occurs mostly in young, unvaccinated dogs. Transmission is primarily through respiratory secretions. Diagnosis of this infection can be difficult and is made based on age of the dog, vaccination status, characteristic symptoms, and by ruling out other diseases.

Severity of symptoms is variable, the earliest of which include sneezing, and ocular and nasal discharge. As the infection progresses, pneumonia, vomiting, diarrhea, and ultimately neurological symptoms can occur. With time, some lucky pups manage to get rid of this virus on their own. Others succumb, especially once neurological symptoms become apparent. Most dogs that are severely affected are euthanized because of the poor prognosis.

Questions to Ask Your Vet

1. **Have we ruled out all other diseases that could be causing these symptoms?**
2. **How long until we know whether or not my dog will improve?**
3. **How will I know if he is getting better or worse?**
4. **What can I be doing at home to support my puppy while his immune system is fighting the infection?**
5. **Have my other dogs been exposed? What can I do to protect them?**

Diagnosis: Fungal disease (coccidioidomycosis, valley fever, histoplasmosis, blastomycosis, cryptococcosis, aspergillosis)

There are several types of *fungal organisms* capable of causing infection in dogs. In various geographic locations, specific types are ubiquitous in the environment. The route of infection is via the respiratory tract. Although inhalation of such organisms occurs frequently, actual infection is rare. Heavy exposure (digging in the soil, exposure to recently excavated soil, hunting with nose close to the ground) and immune suppression caused by disease or medication predispose to infection.

The diagnosis of fungal disease is made by specific blood tests, microscopic evaluation, and fungal culture of infected tissue samples. Symptoms include: decreased appetite; weight loss; and cough. Depending on the type of organism, abnormalities involving the eyes, skin, gastrointestinal tract, kidneys, bones, joints, brain, and spinal cord may develop. Aspergillosis affects the nose and adjacent sinuses (see p. 354) and is often treated by application of a topical anti-fungal solution. Treatment for other types of fungal infections consists of long-term—sometimes lifelong—anti-fungal medication. Prognosis depends on the extent of the infection and how early it was diagnosed and treated.

Questions to Ask Your Vet

1. **Which fungal infection does my dog have?**
2. **Which areas of the body are involved?**
3. **Do you think we are catching the infection in its early stages?**
4. **Is this contagious—can other pets or family members catch it?**
5. **How will we know if and when we've treated my dog long enough?**

* **Don't Forget!** Check p. 283 for Universal Questions, and pp. 148 and 152 for questions when dealing with anesthesia and surgery.

Diagnosis: Leptospirosis

Leptospirosis is a bacterial infection that affects a number of animal species (including humans) and is transmitted via urine. The bacteria can enter the body through skin (especially that which is broken from a cut or scratch) or mucous membranes (eyes, nose, mouth). Dogs that drink free-standing water in an area inhabited by wildlife, or swim or wade in it, are at risk.

When leptospirosis produces illness it causes kidney failure and/or liver disease. Symptoms include: fever; decreased appetite; vomiting; diarrhea; weakness; and excessive thirst. Many infected dogs will show no symptoms, but may become chronic carriers of the disease and shed (release) bacteria in their urine.

Leptospirosis-induced disease is identified with blood tests. Aggressive treatment with intravenous fluids and antibiotics (penicillin, ampicillin, or amoxicillin) provides the best chance for survival. In general, the earlier the disease is detected and the more aggressive the therapy, the better the prognosis. Although many dogs can be cured, some sustain permanent kidney or liver damage. Others do not survive. Vaccines are available that prevent some varieties of leptospirosis (see p. 110).

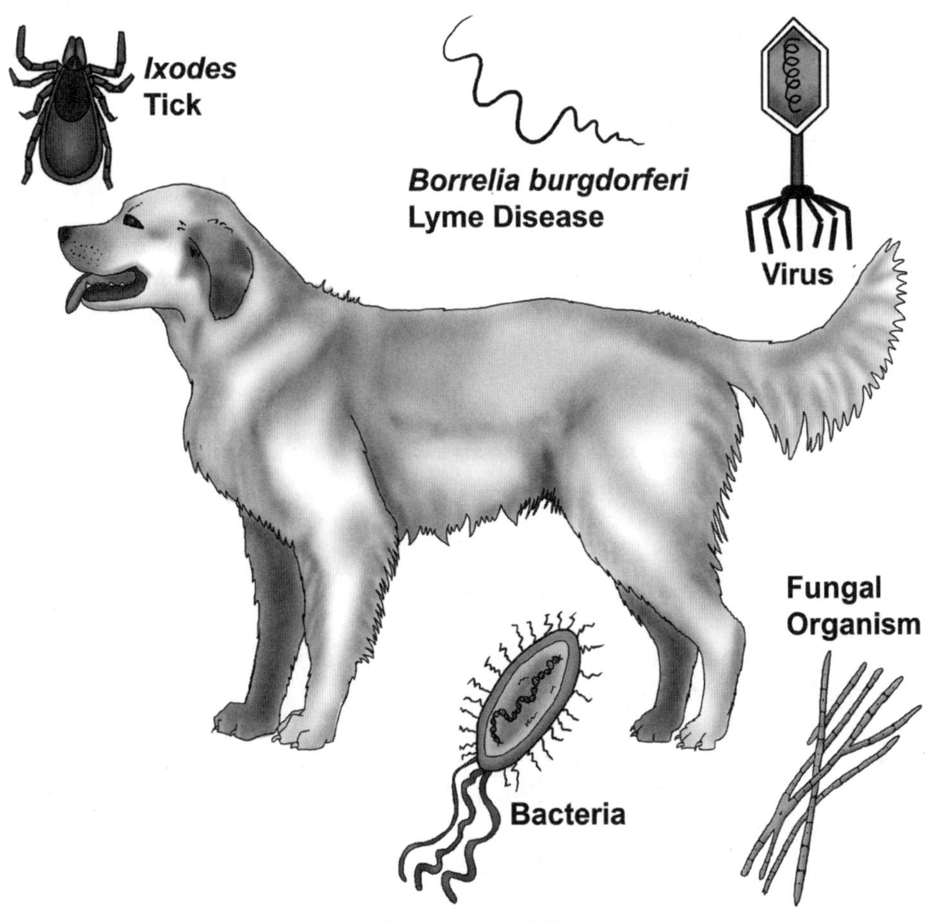

INFECTIOUS DISEASES

Questions for Your Vet

1. Does my dog have liver or kidney involvement? If so, how severe?
2. Did we catch the infection early?
3. Should my other dogs be tested or treated?
4. How long will my dog shed (release) the bacteria in his urine?
5. What do I need to do to avoid getting the infection myself when handling and cleaning up after my dog?
6. How can I prevent my dogs from getting leptospirosis in the future?

Diagnosis: Parvovirus

This is another highly contagious viral infection that affects unvaccinated dogs and occasionally puppies that have not yet completed their initial series of vaccines (see p. 110). *Parvovirus* organisms are present in the stool of an infected dog and can remain infectious in the environment for months. The disease causes protracted and severe vomiting and diarrhea, as well as a marked decrease in white blood cell count, leaving the dog susceptible to secondary bacterial infection.

Parvovirus is diagnosed via a simple test run on a stool sample. Most dogs cannot survive without intensive therapy including intravenous fluids and antibiotics along with medication to control vomiting, diarrhea, low blood sugar, and low protein levels. The prognosis is reasonably good if the dog is treated early and aggressively.

Questions to Ask Your Vet

1. **How sick is my dog (low white blood cell count, low blood sugar, low protein, other parameters)?**
2. **How long do you anticipate he will need to be hospitalized?**
3. **How long will the parvovirus be present in my dog's bowel movements?**
4. **Do I need to keep my other dogs away from him when he comes home?**
5. **What should I do to disinfect our home and yard?**

Diagnosis: Tick-borne diseases (ehrlichiosis, Lyme disease, borreliosis, anaplasmosis, Rocky Mountain Spotted Fever, bartonellosis)

When ticks feed on a dog, they are capable of transmitting a number of serious—even life-threatening rickettsial infections. However, many dogs that are exposed to these organisms never develop any symptoms because their immune systems eliminate the rickettsiae before they do any harm. When dogs do become sick, symptoms include: lethargy; fever; decreased appetite; and joint pain. Identification of the specific disease is made by testing blood or other fluids for the presence of bacteria or antibodies produced in response to the infection. Infected dogs commonly test positive for more than one tick-borne disease because ticks often carry more than one infectious organism.

Antibiotic therapy is prescribed and *tetracycline* drugs are typically the number-one choice. The treatment strategy is also directed toward patient comfort and supportive care. Some dogs require long-term therapy. Prognosis is highly variable and depends on how long the dog has been sick, how severe his symptoms are, and his response to therapy. Some dogs are completely cured while others succumb.

Questions to Ask Your Vet

1. **Which tick-borne disease(s) does my dog have?**
2. **Which organs are affected?**
3. **Does the infection explain all of my dog's symptoms? Can other symptoms arise?**
4. **Can antibiotics alone resolve all my dog's symptoms?**

5. **When should blood tests be repeated?**
6. **How can I safely remove ticks from my dog?**
7. **Which tick-prevention program do you recommend?**
8. **What early symptoms should I be watching for in my other dogs?**

Neuromuscular Diseases

Introduction to the Brain, Spinal Cord, Muscles, and Nerves

The brain is the control center of the body, sending and receiving impulses via the spinal cord and peripheral nerves. Muscles are responsible for converting the neurological signals into action, whether it be the wag of a tail or the churning of intestines.

QUICK REFERENCE

Which Specialist Is Right for My Dog?

Some of these diseases are treated in a general practice setting. When more specialized diagnostics and treatment are desired, your veterinarian will refer you and your dog to a board certified neurologist, or, if not available in your area, to a board certified surgeon, radiologist, or internist.

Diagnosis: Canine cognitive dysfunction (senility)

Canine cognitive dysfunction refers to age-related senility. Symptoms tend to develop gradually and include: disorientation; changes in social behavior (more aloof or more "needy"); "break" in normal house-training; and anxiety-related behavior, such as panting, pacing, barking, and whining. Anxious behavior occurs more commonly during the evening hours.

There is no specific test for canine cognitive dysfunction. Rather, the diagnosis is considered based on a dog's age, typical behavioral changes, and tests that rule out other primary brain diseases. *Selegiline* is a drug licensed for treatment. It helps some dogs and is free of significant side effects. Other treatment options include: behavioral modification; mental and physical stimulation; nutraceuticals (supplements added to the diet); and sedation, especially in the evening when restless behavior is most likely to occur. Diets rich in antioxidants may be of benefit.

Questions for Your Vet

1. **Have other brain diseases been ruled out? If not, what tests can be done?**
2. **How long might it take to see improvement with any of the treatment options?**
3. **How fast will the disease progress?**
4. **What are some parameters I can use to gauge the quality of my dog's life?**

Diagnosis: Caudal cervical spondylomyelopathy (Wobbler syndrome)

The bones that form the spine in the neck region are called *cervical vertebrae*. In dogs diagnosed with *caudal cervical spondylomyelopathy* or *Wobbler syndrome*, they are malformed or out of alignment, causing pressure to be put on the spinal cord. The result is a gradual progression of neck pain and neurological abnormalities that affect all four legs. Although this is a congenital abnormality (a birth defect), symptoms typically don't arise until the dog is middle-aged. Wobbler syndrome occurs primarily in large or giant dog breeds, most commonly Doberman Pinchers.

Wobbler syndrome is best confirmed with an MRI scan. A *myelogram* (X-ray taken with contrast material infused around the spinal cord) may also be diagnostic. Surgery to alleviate the pressure on the spinal cord is often the treatment of choice; most dogs respond favorably, although the recovery process is typically prolonged. Medical therapy consists of rest and medication to control pain and inflammation.

Questions to Ask Your Vet

1. **May I see the study that documented my dog's Wobbler syndrome?**
2. **Is my dog a better candidate for medical or surgical therapy?**
3. **If surgery is recommended, what will the post-operative recovery period be like?**
4. **How can I tell if he is experiencing pain?**
5. **What changes can I make to help him be more comfortable?**
6. **Should I take him to a specialist in rehabilitation therapy?**

Diagnosis: Degenerative myelopathy (German Shepherd dog myelopathy)

This is a most disheartening diagnosis because neither the cause nor any effective therapy is known. It occurs predominantly in German Shepherds and other large breed dogs. Due to a gradual degeneration of the spinal cord, affected dogs experience: gradual weakening of the hind legs; scuffing of the feet and legs; difficulty jumping; loss of bladder and bowel control; and ultimately, paralysis of the hind legs. The front legs are less commonly affected. Fortunately, no pain is associated with the process.

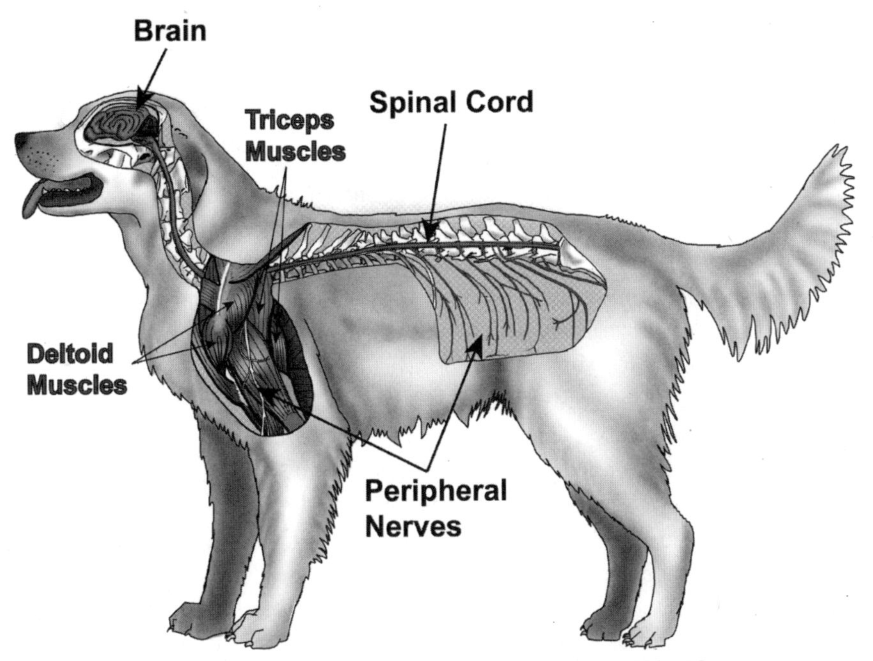

THE BRAIN, SPINAL CORD, MUSCLES, AND NERVES

There is no specific test for degenerative myelopathy. It is diagnosed by ruling out other causes of spinal cord symptoms. Various treatments (vitamin E, vitamin B12, aminocaproic acid) have been discussed, but none have been proven to be effective in slowing the degeneration. Exercise and rehabilitation therapy may help prevent development of secondary muscle atrophy. Nursing care is most important in preventing worn toenails, pressure sores, bladder infections, and scalding of the skin from urine and feces. For some dogs, a wheeled cart that supports the hind end allows them to experience a good quality of life. For most dogs, though, the long-term outlook is bleak.

Questions for Your Vet

1. **Have we ruled out all other causes of spinal cord disease? If not, how can this be done?**
2. **How should I revise my dog's exercise regimen?**
3. **What can I do at home to assist him with his mobility, urination, and bowel movements?**
4. **How can I help him adapt psychologically to his changing abilities?**
5. **Should I decrease his caloric intake to avoid weight gain?**
6. **Is he a candidate for rehabilitation therapy?**
7. **How long do you anticipate it will be until he is no longer able to walk on his own?**

❊ **Don't Forget!** Check p. 283 for Universal Questions, and pp. 148 and 152 for questions when dealing with anesthesia and surgery.

Diagnosis: Epilepsy (idiopathic epilepsy)

Epilepsy is a seizure disorder. "Idiopathic" means no underlying cause has been identified. Diagnosis is based on: early age of onset (one to five years); apparent normalcy between seizures; and exclusion of other causes (toxins, infection, inflammation, tumors, trauma, liver failure, and low blood sugar).

Generalized or *grande mal* seizures are most common. They are characterized by loss of consciousness and generalized involuntary muscle contractions. Milder *focal* or *partial* seizures can also occur that involve only one part of the body, with or without loss of consciousness. The frequency of seizures is variable but tends to increase with age.

Epileptic seizures are typically well controlled with the anti-seizure medications *phenobarbital* and *potassium bromide*, used either singly or in combination. Both can have significant side effects. Dietary salt levels can raise or lower the concentration of potassium bromide in the bloodstream, thereby causing effectiveness or toxicity. Blood levels of both drugs are measured to avoid toxic blood levels or undertreatment. The decision to treat or not treat epilepsy is based on the frequency and severity of epileptic seizures.

Questions for Your Vet

1. **Have we ruled out other causes of seizures? If not, how can this be done?**
2. **Are my dog's seizures frequent or severe enough to warrant treatment?**
3. **What should I do if he has another seizure?**
4. **What constitutes the need for an emergency visit?**
5. **Can my dog do all of his normal activities?**
6. **Will phenobarbital affect any of the other medication my dog is taking?**
7. **If my dog is prescribed potassium bromide, should I discuss diet changes with you in advance?**

Diagnosis: Fibrocartilaginous embolism (fibrocartilaginous embolic myelopathy)

This disease is caused by a clot (*embolism*) in a spinal cord vessel that results in sudden paralysis of one or more legs. Sometimes mild trauma or vigorous exercise occurs right before the onset of symptoms. Although some dogs cry out at the time of onset, most dogs with *fibrocartilaginous embolism (FCE)* seem quite comfortable.

An MRI scan is the test of choice for ruling out other causes of paralysis and revealing changes consistent with FCE. Treatment consists of nursing care, rehabilitation therapy, and patience. Medication and surgery do not play a role. Complete recovery often occurs, but is dependent on the location and severity of the embolism.

Questions to Ask Your Vet

1. **Is it likely my dog will make a complete recovery? How long might this take?**
2. **When do you anticipate he will be able to walk again?**
3. **What type of nursing care should I provide at home?**
4. **Is he a candidate for rehabilitation therapy?**

Diagnosis: Granulomatous meningoencephalomyelitis

This is a noninfectious inflammatory condition that affects the lining of the brain and spinal cord (*meningo*), as well as the brain (*encephalo*) and spinal cord (*myel*) themselves. The cause of *granulomatous meningoencephalomyelitis (GME)* is unknown, but is thought to be immune mediated (the immune system attacks the body's own tissues). It occurs primarily in small dogs. Symptoms vary depending on which parts of the brain and spinal cord are affected, and most commonly include: reluctance to move; neck pain; a head tilt; circling; blindness; and seizures.

GME is diagnosed based on analysis of *cerebrospinal fluid* (fluid surrounding the brain and spinal cord), and brain and spinal cord imaging (MRI or CT scan). It is treated with medications that suppress the immune system, decrease inflammation, manage pain, and control seizures. The response to therapy is highly variable; some dogs will have a good quality of life for a year or more while others will succumb within days to weeks. Treatment is lifelong and long-term prognosis is poor.

Questions to Ask Your Vet

1. **Does the GME involve my dog's brain, spinal cord, or both?**
2. **Is it likely that he will have a seizure? If so, what should I do when it happens?**
3. **Is he in pain? If so, what can I do to increase his comfort (i.e., switch from a neck collar to a harness, elevate food and water bowls)?**
4. **Do I need to limit my dog's activity?**

Don't Forget! Check p. 283 for Universal Questions, and pp. 148 and 152 for questions when dealing with anesthesia and surgery.

Diagnosis: Intervertebral disk disease (slipped disk)

Between the bones of the spine (vertebrae) are cartilaginous shock absorbers called *disks*. Age-related changes or trauma cause the disks to degenerate and protrude into the spinal canal, putting pressure on the spinal cord. *Intervertebral disk disease (IVDD)* can occur as a *chronic* process or an *acute* event. Most dogs experience considerable pain. When present, neurological abnormalities range from mild weakness to complete paralysis. If the IVDD is in the neck, all four limbs may be affected. When located in the back, symptoms may be limited to the hind limbs.

Plain X-rays may suggest IVDD, but an MRI, CT scan, or myelography (X-rays following an injection of contrast around the spinal cord) is necessary for definitive diagnosis. Dogs that have pain, but no neurological deficits (their gait remains normal) are good candidates for medical therapy that might include: rest; pain medication; acupuncture; and anti-inflammatory drugs. If a dog has neurological deficits (gait abnormalities) and/or medical therapy fails, surgery may be needed to relieve the pressure on the spinal cord. The prognosis depends on the degree of neurological impairment and how quickly treatment is provided.

Questions for Your Vet

1. **Can you point out the disk disease on my dog's imaging study?**
2. **Where is the disease located (in which part of the spine)?**
3. **Does my dog have any neurological deficits? If so, which legs are affected?**
4. **Is he a better candidate for medical or surgical treatment?**
5. **How can I assess his comfort level at home?**
6. **What are your recommendations for keeping him strictly confined?**
7. **Once he is more fully recovered, would he benefit from rehabilitation therapy?**
8. **What can I do to prevent another episode of IVDD?**

Diagnosis: Lumbosacral stenosis (cauda equina syndrome)

The lumbosacral region refers to the area of the spine over the hips. Age-related degenerative changes here can cause a narrowing (*stenosis*) of the spinal canal resulting in compression of nerve roots. Affected dogs may exhibit pain when the tail is manipulated (moved up and down or from side to side) or manual pressure is applied over this region of the back. They may also exhibit: hind leg weakness/dragging; urinary/fecal incontinence; and reluctance to jump, rise, and climb stairs.

X-rays can suggest the presence of the disease, but MRI and CT scans tend to be much more definitive. Medical management of lumbosacral stenosis includes rest, anti-inflammatory medication, and pain control. Surgery is recommended for dogs with refractory pain (pain that isn't improving in response to medical treatment and time) or significant neurological dysfunction. The goal of surgery is "decompression" of the affected nerve roots. Prognosis is variable and tends to be a bit worse for dogs with urinary or fecal incontinence.

Questions to Ask Your Vet

1. **May I see the imaging study of my dog's lumbosacral region?**
2. **Is he a better candidate for medical or surgical therapy?**
3. **How long will he need to be confined after surgery?**
4. **What is the best way to gauge his pain level?**
5. **Will rehabilitation therapy be helpful?**
6. **What is the likelihood that my dog's symptoms will fully resolve?**

Diagnosis: Myasthenia gravis

Myasthenia gravis is a disease that interferes with normal muscle activity. A muscle normally contracts when a nerve impulse triggers receptors on the muscle's cells. When a dog has myasthenia, these receptors are in short supply, destroyed by the body's own immune system. Disease involvement of skeletal muscles causes profound weakness and fatigue. Affected dogs are able to take a few steps, then weaken or collapse. The muscles of the esophagus can be affected, resulting in flaccid, nonfunctional *megaesophagus* (see p. 309), and regurgitation of food and water, which can then be inhaled into the lungs, resulting in *aspiration pneumonia*.

The most definitive test for diagnosing myasthenia is the *anti-acetylcholine (ACH) receptor antibody test*. It is a blood test that measures immune system activity against the muscle receptors. A less precise test involves administration of *tensilon*, a drug that temporarily increases stimulation of the existing receptors. Positive results of tensilon testing can be dramatic—a dog unable to walk is suddenly capable of trotting down the hallway, though this effect lasts for only a minute or two. Even if the tensilon test is negative (no apparent change is detected), myasthenia should still be fully ruled out by way of the antibody blood test. Chest X-rays are used to document whether or not the esophagus is affected.

Pyridostigmine, a drug that maximizes the activity of the remaining muscle receptors, is the treatment of choice in most cases of myasthenia gravis. Medications that suppress the immune system may also be recommended. Treatment specific for megaesophagus involves changes in feeding habits and treatment of any secondary aspiration pneumonia. In general, esophageal involvement means a poorer prognosis because of the potential for life-threatening pneumonia. Assuming they survive the initial stages of the disease, many dogs have spontaneous remission after months or years of treatment.

Questions to Ask Your Vet

1. **Does my dog have esophageal involvement, skeletal muscle involvement, or both?**
2. **Is there any evidence of aspiration pneumonia?**
3. **What are the symptoms of aspiration pneumonia? If seen, should they prompt an emergency visit?**
4. **Is there a need to change my dog's feeding regimen to prevent regurgitation?**
5. **When might we expect to see improvement?**

Diagnosis: Vestibular disease

The *vestibular system* is responsible for normal balance. Part of this system is located within the inner ear, and when disease occurs here, it is referred to as *peripheral* vestibular disease. *Central* vestibular disease refers to an abnormality within a particular region of the brainstem. Both central and peripheral vestibular diseases cause a dog to feel terribly dizzy. Symptoms can include: a head tilt; *nystagmus* (the eyes dart quickly back and forth); nausea; a staggering, somewhat drunk-appearing gait; and even whole-body rolling. In cases of central vestibular disease, there may also be decreased mental responsiveness and leg weakness.

The diagnosis of vestibular disease is based on: specific neurological symptoms; screening for infectious diseases; brain imaging (MRI or CT scan); and analysis of cerebrospinal fluid (the fluid that surrounds the brain and spinal cord). It is important to differentiate between peripheral or central vestibular disease.

The most common cause of peripheral vestibular disease is *idiopathic vestibular syndrome*. Thought to be a stroke-like event affecting the inner ear of older dogs, it results in an abrupt onset of severe symptoms. Euthanasia is often contemplated by upset family members; however, with good nursing care, most dogs make a dramatic recovery within days. Inner ear infections also cause peripheral vestibular disease and can be treated with antibiotics and, sometimes, surgical drainage. The prognosis is good in most cases of peripheral vestibular disease.

Infectious, inflammatory, and cancerous diseases are the most common causes of central vestibular disease. By virtue of its location in the brain, central vestibular disease carries a more guarded prognosis.

Questions to Ask Your Vet

1. **Does my dog have central or peripheral vestibular disease?**
2. **If he has peripheral disease, have we ruled out an ear infection as the underlying cause?**
3. **How long do you anticipate it will be until he is better?**
4. **If he has central vestibular disease, do we know what the underlying cause is? If not, how can we find out?**
5. **What type of nursing care is needed if I take my dog home?**

Ocular Diseases

Introduction to the Eye

What a complex little organ the eye is—so many parts must function properly in order for normal vision to occur. The *cornea* is the clear surface layer of the eye, the *iris* is the colored part of the eye that controls pupil size, and the *lens* sits directly behind the iris. In its normal state, the lens is crystal clear (the reason the pupil appears black). It focuses light on the retina, the sensory portion of the eye. The retinas—each no larger than the peel of a grape yet capable of transmitting a visual feast to the brain.

QUICK REFERENCE

Which Specialist Is Right for My Dog?

When extra expertise is needed with canine ocular diseases, you and your family veterinarian will want to solicit help from a board certified veterinary ophthalmologist.

Diagnosis: Cataracts

In a normal, healthy dog, the pupil of the eye appears black because the lens is crystal clear. When the pupils become gray or white, it is because he has *cataracts* or a normal aging condition called *lenticular sclerosis*. Whereas a cataract can cause blindness by blocking transmission of light to the retina, lenticular sclerosis causes no significant visual impairment and requires no treatment. A simple examination can discriminate between cataracts and lenticular sclerosis.

There is a genetic predisposition to cataract formation. *Diabetes mellitus* and diseases within the eye (inflammation, glaucoma, lens luxation) can also cause cataracts.

They usually form gradually, giving the dog time to adapt to his vision loss. The decision to remove the cataracts to restore vision should only be made after thorough discussion with a veterinary ophthalmologist regarding the risks and benefits of surgery. Keep in mind, if surgery is not feasible, a blind dog can have an exceptionally good quality of life. When you opt for surgery, the ophthalmologist will perform some noninvasive tests to be sure that cataract removal will restore vision. Even if you are not considering surgery, or your dog's cataracts are "immature" (not yet causing a visual disturbance), a trip to the ophthalmologist is warranted to diagnose and treat any concurrent eye disease.

Questions to Ask Your Vet

1. **Do we know what caused the cataracts?**
2. **Does my dog have any remaining vision?**
3. **If I opt for surgery, will it be performed on one or both eyes?**
4. **If surgery is not an option, what can I do to enhance my dog's quality of life and ensure his safety?**
5. **Are the cataracts causing any other eye problems, such as glaucoma or inflammation within the eye?**

Diagnosis: Conjunctivitis

Conjunctivitis refers to inflammation of the *conjunctiva*, the pink tissue that lines the inner surface of

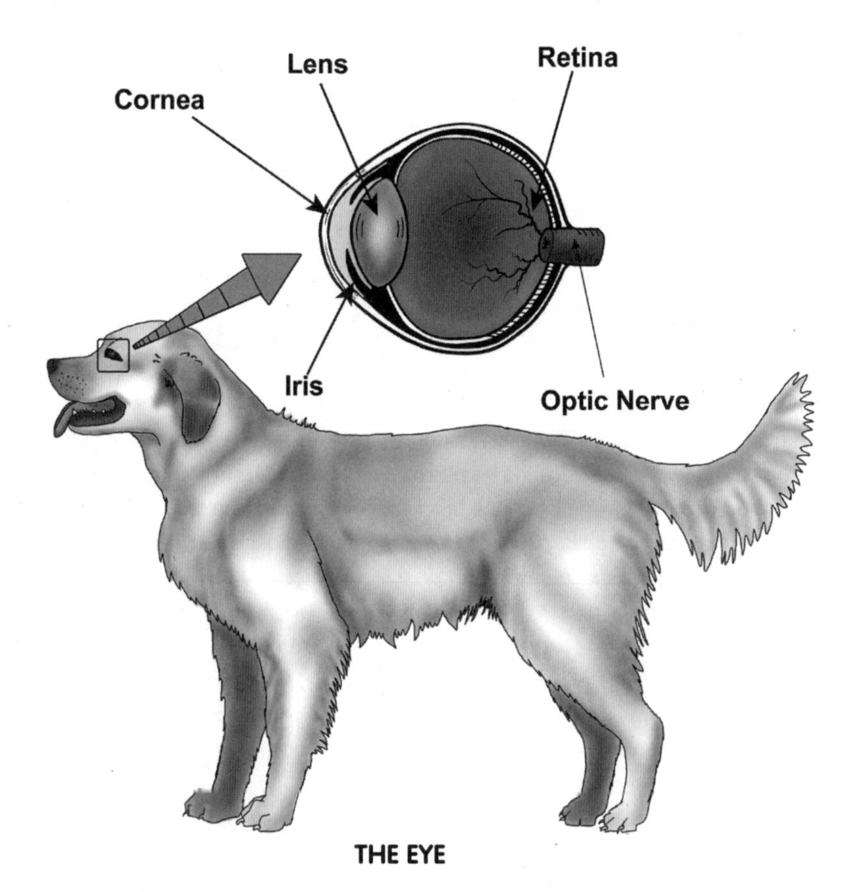

THE EYE

the eyelids. Symptoms are often redness and discharge. Causes include: allergies; environmental irritants; infection; eyelid abnormalities; and inflammation of the skin around the eyes. Conjunctivitis often accompanies other more serious eye diseases, such as glaucoma, "dry eye," and corneal ulcers.

Treatment is directed at the underlying cause. *Allergic* conjunctivitis is typically treated with topical cortisone and *bacterial* conjunctivitis with topical antibiotics. The prognosis is usually good.

Questions for Your Vet

1. **What is the cause of my dog's conjunctivitis? Is it contagious to my other dogs?**
2. **Are one or both eyes affected?**
3. **Does he have any other eye abnormalities? If so, how will they be treated?**
4. **Is it likely that he will need long-term treatment?**
5. **Do I need to prevent him from rubbing his eyes?**

Diagnosis: Corneal ulcer (corneal abrasion)

The cornea is the perfectly clear surface layer of the eye, and an *ulcer* is a scratch or abrasion of this surface. Trauma, decreased tear production, infection, eyelid defects, and facial conformation ("smoosh-faced" breeds, such as Pugs and Boston Terriers, are at risk) are causes of corneal ulceration. Symptoms include: squinting; redness and swelling around the eye; and ocular discharge. Corneal ulcers are diagnosed by placing fluorescein dye into the cornea, which "stains" the ulcer (making it easy to evaluate).

Not only are corneal ulcers painful, they can lead to scarring of the cornea and infection within the eye itself. Treatment of a *superficial* corneal ulcer includes pain management, antibiotics, and use of an Elizabethan collar to prevent self-trauma. An ulcer is referred to as *indolent* when it fails to heal in response to conventional therapy (a common problem in Boxers). Indolent ulcers, or those that are deep, require more advanced therapy (some of which is surgical). Most family veterinarians are adept at treating superficial ulcers, but those that involve deeper layers of the cornea, and especially those that are indolent, are commonly referred to a veterinary ophthalmologist.

Questions to Ask Your Vet

1. **Do we know what caused my dog's ulcer? If so, is it something that requires specific therapy?**
2. **Are one or both eyes affected?**
3. **What should I do if I'm having trouble administering the eye medication?**
4. **When can we expect corneal healing to be complete?**
5. **Is the ulcer causing secondary problems such as inflammation within the eye?**
6. **What is the likelihood for recurrence?**

Diagnosis: Eyelid abnormalities (entropion, ectropion, distichiasis, ectopic cilia, tumors)

Eyelids can have a variety of abnormalities. With *entropion*, the lids "curl in," causing the eyelashes to rub against the cornea (visualize the way a Shar-Pei appears). Bloodhounds commonly have *ectropion*—the lower eyelid "curls outward," creating exposure and inflammation of the *conjunctiva* (the pink tissue on the underside of the lid). Other eyelid abnormalities include *distichiasis* (an extra row of eyelashes on the inner eyelid margin), *ectopic cilia* (individual eyelashes that grow inward rather than outward), and tumors that can be benign or malignant. Such conditions can cause chronic irritation to the corneal surface or the conjunctiva (the pink tissue that lines the inner surface of the eyelids).

Treatment of eyelid abnormalities consists of removal or correction of the defect. Entropion and ectropion surgeries are considered "plastic surgeries" and are best performed by an experienced surgeon. There are several ways of treating abnormal eyelashes, including electrolysis, *cyrotherapy* (freezing the hair follicles), and *laser ablation* (destroying the follicles with a laser beam). Prognosis depends on the skill of the veterinarian correcting the defect.

Questions to Ask Your Vet

1. **Is treatment necessary?**
2. **Does my dog currently have any other eye problems? If so, how should they be treated?**
3. **What is the likelihood that the problem will be permanently corrected?**
4. **Is this an inherited defect? Should I abandon plans to breed my dog?**

✻ **Don't Forget!** Check p. 283 for Universal Questions, and pp. 148 and 152 for questions when dealing with anesthesia and surgery.

Diagnosis: Glaucoma

Glaucoma is increased pressure (fluid buildup) within the eyeball. Symptoms include: discomfort; corneal clouding; dilated pupils; and redness of the eye. Chronic glaucoma produces enlargement of the eyeball. Glaucoma can cause blindness so when it comes to diagnosis and treatment, time is of the essence.

Certain breeds are predisposed to *primary* glaucoma (Basset Hounds, Chow Chows, American Cocker Spaniels, Shar-Peis, and Labrador Retrievers lead the pack). *Lens luxation* (the lens slips forward from its normal position behind the pupil) and infectious and inflammatory conditions of the eye can produce *secondary* glaucoma. Diagnosis is made using a *tonometer*, a device that measures intraocular pressure. The mainstay of therapy includes medication to decrease pressure and treatment of any underlying ocular disease. Surgical procedures are also available. In some cases, *enucleation* (removal of the eye) is the best solution when blindness is permanent and the eye is painful.

Questions to Ask Your Vet

1. **Do we know what caused my dog's glaucoma? If so, should it be treated in a specific fashion?**
2. **Are both eyes affected? If the second eye is not affected, what is the likelihood that it will also develop glaucoma?**
3. **What is the prognosis for vision in the affected eye?**
4. **Do you sense that my dog is experiencing pain from the glaucoma?**

Diagnosis: Keratoconjunctivitis sicca (dry eye)

Keratoconjunctivitis sicca (KCS) refers to inflammation of the cornea (clear surface of the eye) and conjunctiva (the inside layer of the eyelids) caused by decreased tear production and consequent "dry eye." There is a definite breed predisposition (Cocker Spaniels, Shih Tzus, Lhasa Apso, Pugs, English Bulldogs, West Highland White Terriers, Miniature Schnauzers, Miniature Poodles, and Pekingese top the list). KCS is most commonly caused by destruction of the *lacrimal* (tear-producing) glands by the body's own immune system. Hormonal imbalances and neurological diseases can also cause KCS. Dry eye can occur as an adverse drug reaction to a handful of medications used to treat other canine diseases.

KCS symptoms include: red eyes; ocular discomfort; and thick green or yellow discharge. It can result in *secondary corneal ulceration* (see p. 335). The disease is diagnosed by inserting a narrow strip of filter paper under the eyelid to measure tear production. Left untreated KCS results in permanent scarring of the corneal surface and blindness, not to mention chronic discomfort. Lifelong treatment consisting of frequent cleaning of the eyes and topical application of lubricants and medication to help boost tear production is usually necessary. Sometimes, surgery is performed, transplanting a duct that normally transports saliva into the mouth under the eyelid. The saliva provides a source of constant lubrication for the eye—even more so when the dog thinks he's about to be fed!

Questions to Ask Your Vet

1. **Do we know what caused my dog's KCS? If so, should it be treated in a specific way?**
2. **Are both eyes affected?**
3. **Have the corneas been damaged?**
4. **What should I do if I'm having difficulty administering medication?**
5. **Do I need to stick to a rigid schedule when applying medication?**

Diagnosis: Lens luxation

The lens is located just behind the pupil. Its purpose is to focus light on the retina, the sensory portion of the eye. A *luxated lens* has moved out of position (dislocated) forward in front of the iris, or backward too far behind it. Lens luxation occurs as an inherited defect, or can arise secondary to other eye disease, such as trauma, glaucoma, cataracts, uveitis, or cancer (see p. 332 for the directory to these specific diseases).

Treatment consists of surgical removal of the luxated lens and treatment of any underlying condition. The surgery is highly technical and performed by a veterinary ophthalmologist. In general, the earlier surgery is performed, the better the prognosis.

Questions to Ask Your Vet

1. **Are both lenses luxated or only one?**
2. **Is the lens luxated forward or backward within the eye?**
3. **Has an underlying cause been identified? If so, how will it be treated?**
4. **What are the potential complications associated with surgery?**
5. **What is the likelihood that vision will be preserved?**
6. **Is the lens luxation causing other eye abnormalities such as glaucoma or inflammation within the eye?**

Diagnosis: Ocular trauma/foreign body

Dogs get into all kinds of mischief. Sometimes it's their eyes that pay the price, from plant material lodged under the eyelid, to getting sprayed in the face by a skunk, to diving into a thornbush headfirst looking for a tennis ball. Some smoosh-faced breeds (Pugs, Pekingese, Shih Tzus) have shallow eye sockets and are prone to *ocular proptosis*, where the eyeball is displaced from the socket when certain types of pressure or trauma occur around the eye. Anytime the eye is traumatized, there is risk of vision loss and urgent care is a must. This isn't something that can wait until Monday morning. Treatment is aimed at providing comfort, minimizing inflammation, and preserving normal vision.

Questions to Ask Your Vet

1. **What damage has the eye sustained?**
2. **Is there damage to other facial structures?**
3. **Is it likely that my dog will sustain permanent vision damage?**
4. **Is surgery required?**

Diagnosis: Retinal detachment

The retina is the sensory nerve layer in the back of the eye that perceives an image and then transports it via the optic nerve to the brain. *Retinal detachment* is a separation of the layers of the retina, causing loss of visual perception at the affected site(s). Retinal detachments may be an inherited disorder, or they may be caused by trauma, high blood pressure, blood clotting abnormalities and other ocular diseases (glaucoma, lens luxation, uveitis, cataracts). Unless both eyes are significantly affected and there is substantial visual impairment, there may be no obvious symptoms. Often, retinal detachment is picked up as an incidental finding during routine examination.

Therapy for retinal detachment is aimed at treating the underlying cause. The prognosis for healing is variable and depends on the extent and chronicity of the detachment.

Questions to Ask Your Vet

1. **What is the cause of the detachment?**
2. **What is its extent? Are one or both eyes affected? Can my dog still see out of both eyes?**
3. **Are there any other eye abnormalities present (uveitis, glaucoma, lens luxation, cataracts)?**

Diagnosis: Uveitis

Uveitis refers to an inflammation inside the eye itself. Symptoms include: squinting; red eyes; small pupil size; haziness within the eyeball; and color change of the iris. If there is no apparent ocular cause for the uveitis (corneal ulcer, foreign body, trauma), a search for an underlying infectious, inflammatory, or cancerous cause elsewhere in the body is in order. Unfortunately, even with a full barrage of diagnostic testing, no apparent cause is found in approximately 50 percent of dogs with uveitis.

Uveitis can cause transient or permanent glaucoma, and decreased vision or blindness. Therefore, it should be investigated and treated aggressively. Anti-inflammatory medications are commonly used. Prognosis depends on the underlying cause of the intraocular inflammation.

Questions to Ask Your Vet

1. **Do we know what caused my dog's uveitis? If not, what can be done to find out?**
2. **Are one or both eyes affected?**
3. **Has glaucoma developed as a result?**
4. **Is my dog in pain? If so, how can he be treated?**

❊ **Don't Forget!** Check p. 283 for Universal Questions, and pp. 148 and 152 for questions when dealing with anesthesia and surgery.

Orthopedic Diseases

Introduction to the Skeletal System

The average dog skeleton contains 319 bones. A *joint* is the area where two bones are attached or come together for the purpose of motion. *Orthopedic disease* refers to abnormalities within a bone or joint.

QUICK REFERENCE

Which Specialist Is Right for My Dog?

Dogs with orthopedic diseases are commonly referred to board certified veterinary surgeons (whether or not surgery is needed), rehabilitation therapists, and veterinarians specializing in acupuncture.

Diagnosis: Cruciate ligament injury

There are two cruciate ligaments that stabilize each knee joint. Spontaneous rupture (tearing) of the cranial (anterior) cruciate ligament is relatively common, especially in larger dogs, whose conformation tends to put undue stress on this ligament. The caudal (posterior) ligament is affected less commonly. Tears can be partial or complete, but both result in instability and pain in the stifle (knee).

The most common symptom associated with cruciate ligament rupture is a sudden onset of lameness. The diagnosis is usually made during a physical exam by detecting a characteristic instability of the knee called a "cranial drawer sign." X-rays are commonly recommended to detect any preexisting arthritis or other abnormalities in the joint and take preoperative measurements (depending on the surgical procedure).

Surgery is the treatment of choice for dogs with cruciate ligament damage. It restores stability to the joint making it more comfortable and functional and minimizes the degree of future arthritis. Several different surgical procedures are available. At the time of writing, the two favored are the *TPLO (tibial plateau leveling osteotomy)* and *TTA (tibial tubercle advancement)*. In addition to surgery, weight control, temporary exercise restriction, and nonsteroidal anti-inflammatory medication are recommended. Postoperative rehabilitation therapy (the equivalent of human physical therapy) hastens the recovery process. Unfortunately, even with surgery, approximately 50 percent of dogs go on to tear the cranial cruciate ligament in the opposite leg.

Questions to Ask Your Vet

1. **Does my dog have a partial or complete tear?**
2. **Which surgical procedure do you recommend and why?**
3. **Is the other knee affected?**
4. **What will be the duration and degree of exercise restriction following surgery?**
5. **Will my dog ultimately be able to return to his normal level of activity?**
6. **Should he be taking supplements or nutraceuticals (fish oil, glucosamine, chondroitin sulfate, MSM)?**
7. **Does my dog need to lose weight?**

Diagnosis: Fracture disease

A *fracture* is a broken bone. Trauma is the most common cause of fractures, but occasionally, pathologic fractures result from disease, most commonly cancer, within the bone. Fractures are described based on their orientation in the bone (transverse, oblique, spiral); whether the bone is in two or more pieces (simple or complex); and whether or not the bone has been exposed to the outside world through a break in the skin surface (open versus closed). The nature and location of the fracture determine how best to repair it.

Fractures are often apparent during a physical examination and X-rays are used to characterize the fracture and help determine the most appropriate repair. CT or MRI scans are sometimes recommended when X-rays do not allow complete visualization of the fracture. Fractures are typically treated with *internal* fixation (a stabilization apparatus is surgically applied directly to the bone) or *external* fixation (a cast or splint). When fractures are minimally displaced or don't involve a bone that moves or bears weight (the skull, for example), repair may not be necessary. Antibiotics are warranted with compound fractures. Pain medication is always indicated.

Questions for Your Vet

1. **Can you show me the fracture on my dog's X-rays or other imaging study?**
2. **Was the skin broken when the fracture occurred?**

3. **Are there other injuries besides the fracture? If so, how should they be addressed?**
4. **If the fracture occurred spontaneously (without trauma) what is the underlying disease within the bone?**
5. **How will my dog's lifestyle (and, therefore my lifestyle) need to be modified during the healing process?**
6. **Is my dog a candidate for rehabilitation therapy?**
7. **How will I be able to assess his level of pain?**

Diagnosis: Hip dysplasia

The hip is a ball-and-socket joint: the ball is the top of the femur (thighbone) and the socket (the *acetabulum*) is within the pelvis. *Hip dysplasia* is a laxity of the joint that creates instability. The body responds to this laxity by remodeling the joint surface with extra bone in an attempt to increase stability. The result of this remodeling is *arthritis*.

Hip dysplasia is common in large breed dogs. Genetics play a big role and we've learned that the hip health of the aunts and uncles as well as the parents impacts whether or not a puppy will have hip dysplasia. Nutrition also matters. Feeding high-energy, high-calorie diets to young, large breed dogs causes rapid growth and predisposes them to hip dysplasia.

Most affected dogs don't develop symptoms of hip dysplasia until a few years of age, after arthritis has developed. Symptoms typically include difficulty getting up, stiffness during or following exercise, and a change in jumping behavior. Hip dysplasia is diagnosed by physical examination and X-rays of the hips.

When hip dysplasia is detected before a dog has fully matured, surgery can sometimes be performed in the hopes of preventing future arthritis. Once arthritis has developed, the mainstay of therapy is treatment with: joint health supplements (glucosamine, chondroitin sulfate, MSM, fish oil); acupuncture; weight management; rehabilitation therapy (the equivalent of physical therapy for people); and nonsteroidal anti-inflammatory medications. Cortisone, especially when given long-term, is a treatment of last resort because of the potential for negative side effects. Hip replacement is available for those dogs that fail to respond to medical therapy. With appropriate treatment, most dogs can continue to be active and pain-free for many years.

Questions to Ask Your Vet

1. **Can you show me the abnormalities on my dog's X-rays (keep in mind, the degree of change on the X-rays doesn't always correlate with the degree of symptoms)?**
2. **Are one or both hip joints affected?**
3. **How can I accurately assess whether or not my dog is uncomfortable?**
4. **Does my dog need to lose weight?**
5. **Should his exercise be restricted?**
6. **Is my dog a candidate for surgery?**

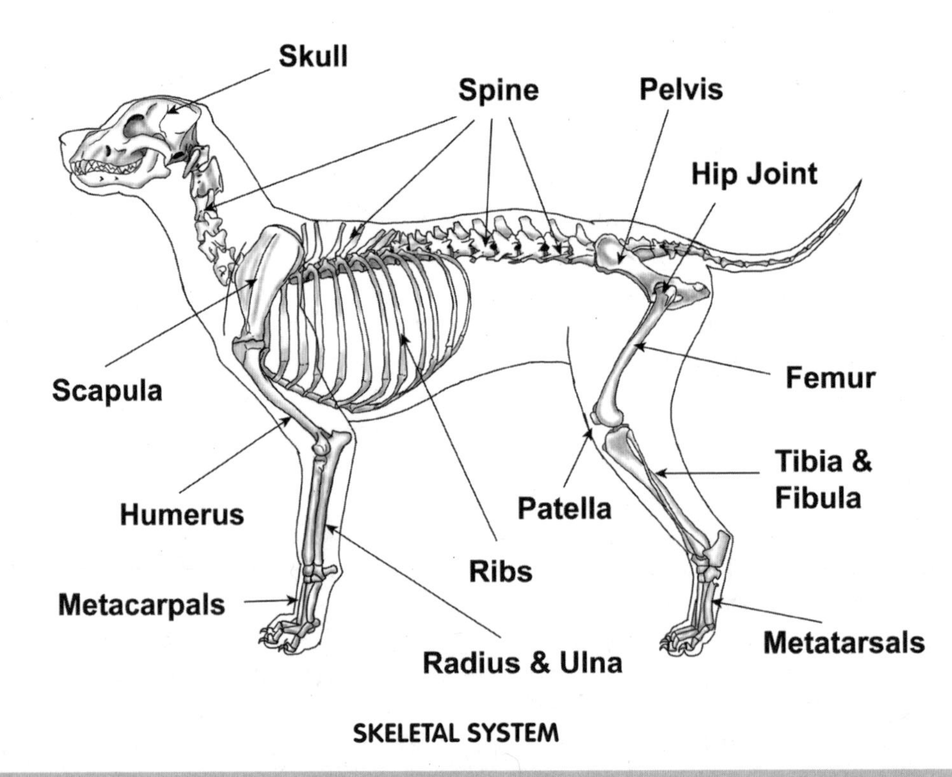

SKELETAL SYSTEM

❊ **Don't Forget!** Check p. 283 for Universal Questions, and pp. 148 and 152 for questions when dealing with anesthesia and surgery.

Diagnosis: Immune mediated polyarthritis (polyarthritis)

Immune mediated polyarthritis (inflammation within multiple joints) is different from the usual wear-and-tear *osteoarthritis* that we anticipate seeing in older dogs (see p. 343). Rather than affecting the cartilage within the joints, immune mediated polyarthritis causes inflammation within the lining (*synovial membrane*) of the joints. Occasionally, an infectious or cancerous process is discovered that has triggered the immune system to behave badly, but usually an underlying cause is not identified. Polyarthritis can occur in conjunction with other immune mediated diseases affecting blood cells and the kidneys.

Symptoms include: difficulty getting up; swollen painful joints; and a "walking-on-eggshells" appearance. Diagnosis is made via microscopic evaluation of joint fluid and testing to rule out infectious causes (bacterial, fungal, and tick-borne diseases). Treatment consists of giving medication to suppress the immune system. Unfortunately, return-to-normal gait and comfort does not guarantee complete resolution of the joint inflammation. It's easy to discontinue medication too early, resulting in return of the polyarthritic symptoms. Follow-up joint fluid analysis is the ideal tool to guide therapy. Most dogs are treated for a minimum of two to three months. Others require even longer term, if not lifelong, therapy. Approximately 30 percent of dogs with immune mediated polyarthritis experience a relapse.

Questions to Ask Your Vet

1. **Which joints are involved?**
2. **Have we ruled out infectious causes of polyarthritis? Are other causes suspected?**
3. **Is there evidence of immune mediated disease in any other body systems?**
4. **When will the joint fluid analysis be rechecked?**
5. **When can my dog resume his normal exercise regimen?**
6. **Should future vaccinations be reconsidered and, perhaps, discontinued for fear of "retriggering" the immune system?**

Diagnosis: Osteoarthritis (arthritis)

Just like people, dogs get *arthritis*—a progressive, irreversible deterioration of cartilage within the joints. It has a propensity for occurring in older, large breed dogs, although any dog can be affected. *Hip dysplasia*—an inherited instability within the hip joint—is the most common cause of osteoarthritis in dogs (over time, joint instability results in arthritis). Other commonly affected sites include the back and elbows. Arthritis can often be detected during a physical examination, but X-rays are used to confirm the diagnosis. Pain is the major symptom, although some dogs show no evidence of discomfort. Activity and cold often exacerbate arthritic symptoms.

The goals of treatment are controlling pain, improving function, and slowing disease progression. Effective therapy often involves using more than one treatment modality and includes: weight management; supplements to support cartilage health (administered orally or injected directly into the joint); prescription diets; acupuncture; rehabilitation therapy; nonsteroidal anti-inflammatory medication; and surgery to either fuse (render it immobile) or replace the joint.

Questions to Ask Your Vet

1. **Which joint(s) is affected?**
2. **Can you show me the X-ray changes?**
3. **How can I best assess if my dog is experiencing discomfort?**
4. **Should I change my dog's diet? Does my dog need to lose weight?**
5. **Does his activity level need to be restricted? If so, to what degree?**

Diagnosis: Osteochondrosis (osteochondritis dessicans, fragmented coronoid process, ununited anconeal process, elbow dysplasia)

Osteochondrosis (OCD) is a defect in the normal maturation of cartilage and bone within a joint. Areas of cartilage thickening develop that disrupt the normal smooth joint surfaces. Sometimes a piece of cartilage breaks away from the surface to form a "joint mouse." Osteochondrosis occurs most commonly in the shoulder, elbow, stifle (knee), and hock (ankle) joints. *Fragmented coronoid process* and *ununited anconeal process* are two forms of OCD that affect different regions of the elbow joint. Genetic and nutritional factors (high-calorie, high-energy diets resulting in rapid growth) predispose dogs to OCD. Large and giant breed dogs are most commonly affected, and males more than females, perhaps due to their larger stature.

Symptoms first appear in dogs under a year of age. Lameness, especially following exercise, is commonly observed. Symptoms can wax and wane when a "joint mouse" is present. Osteochondrosis can often be diagnosed with X-rays alone. Other diagnostic options include MRI and CT scans, as well as *arthroscopy* (a telescopic device is inserted into the joint through a small incision to directly visualize the cartilage surfaces). Surgical treatment of OCD (either via traditional surgery or arthrosocopy) is used to smooth the cartilage surface, remove any "joint mice," and treat the underlying bone to maximize healing and normal growth of its cartilage surface. Nonsurgical treatment options include: weight management; supplements to support cartilage health (administered orally or injected directly into the joint); prescription diets; rehabilitation therapy; and nonsteroidal anti-inflammatory medication. Left untreated, OCD results in arthritis and chronic discomfort.

Questions to Ask Your Vet

1. **Which joint(s) is affected? How severely?**
2. **Can you show me the images that demonstrate the OCD?**
3. **Should I change my dog's diet?**
4. **Does he need to lose weight?**
5. **How can I gauge his comfort level?**
6. **Should I change his activity level?**

Diagnosis: Patellar luxation (medial patellar luxation, slipped knee cap)

The *patella* (knee cap) moves smoothly up and down within a bony groove when the knee is flexed and extended. Conformation abnormalities in some dogs result in the kneecap popping in and out of the groove during motion of the knee. *Medial* patellar luxation (the patella shifts toward the inside of the knee) is common in toy and miniature breeds. *Lateral* patellar luxation (the patella shifts toward the outside of the knee) is rare.

The most common symptom is intermittent non-weight bearing lameness—the dog has a transient hiking-up of the leg that looks like a skipping gait when walking or running. Diagnosis is made by feeling the patella dislocate out of the groove during manipulation of the leg in conjunction with ruling out other causes of the lameness.

Treatment options include medical management with nonsteroidal anti-inflammatory medication, rehabilitation (physical) therapy, and surgery to repair the conformation defects in the knee. Typically, the more severe the luxation, the more aggressive the treatment plan needs to be. Minimally affected dogs require no treatment. The prognosis is good with appropriate therapy.

Questions for Your Vet

1. **Are the symptoms severe enough to warrant treatment? Are they bad enough to require surgery?**
2. **Are both knees affected? If so, is one worse than the other?**
3. **How painful is this condition?**
4. **Should my dog's activity level be restricted?**
5. **Does he need to lose weight?**
6. **Should he be taking supplements or nutraceuticals (fish oil, glucosamine, chondroitin sulfate, MSM)?**

❊ **Don't Forget!** Check p. 283 for Universal Questions, and pp. 148 and 152 for questions when dealing with anesthesia and surgery.

Reproductive Diseases

Introduction to the Reproductive System

The male reproductive system consists of the *testicles, prostate gland,* and *penis*. Both testicles are removed when castration is performed. The *ovaries, uterus, cervix,* and *vagina* comprise the female reproductive tract. Spaying (ovariohysterectomy) is the surgical removal of the uterus and both ovaries.

QUICK REFERENCE

Which Specialist Is Right for My Dog?

If extra help is needed with your dog's reproductive disease, your veterinarian will refer you to a specialist in reproductive or internal medicine.

Diagnosis: Cryptorchidism (retained testicles)

During fetal development, the testicles mature in the abdomen and gradually descend down the *inguinal canal* (the crease between the inside of the leg and the belly) into the scrotum. Both testicles should be present within the scrotum at birth or shortly thereafter. Failure of descent of one or both testicles is called *cryptorchidism*, which means "hidden testicle." Because it is primarily an inherited condition, affected dogs should not be used for breeding. Retained testicles have a 10-fold increased risk for testicular cancer. Removal of *both* testicles is the treatment of choice. If the surgeon is uncertain of the location of the "missing" testicle, presurgical ultrasound imaging might be recommended.

Questions to Ask Your Vet

1. **How long should we wait to be truly certain that the retained testicle(s) isn't going to descend?**
2. **How can we be sure my dog isn't actually missing a testicle(s)?**
3. **How will the recovery process differ from that associated with routine castration surgery?**

Diagnosis: Dystocia (ineffective labor)

Dystocia refers to difficulty in the normal vaginal delivery of a puppy from the uterus. Causes include: oversized pups (occurs more commonly with small litters); structural and conformational abnormalities of the uterus or birth canal; and maternal health issues. Brachycephalic breeds ("smoosh-faced" dogs, such as Pugs, Boston Terriers, Shih Tzus) are particularly predisposed to dystocia because the pup's head size is large in comparison to the birth canal.

Dystocia is suspected when there is a deviation from the normal sequence of events (labor not progressing, too much time between pups, stillbirths, abnormal-appearing vaginal discharge). Diagnosis is confirmed based on: history; a thorough physical examination; imaging with X-rays or ultrasound; and, if available, uterine monitoring.

Treatment of dystocia may be medical, surgical, or a combination of the two. A drug callled oxytocin (increases the frequency of uterine contractions) and calcium (increases the strength of uterine contractions) are commonly administered. Traction may be used to remove a puppy that is stuck within the birth canal. When surgery is necessary, a Cesarean section is performed. Removal of the uterus is strongly recommended if subsequent breeding is not a priority. With appropriate timing and therapy, the prognosis for the dam is good. The prognosis for the puppies is fair to good.

Questions to Ask Your Vet

1. **What is the cause of the dystocia?**
2. **Is my dog a better candidate for medical or surgical treatment?**
3. **Will nursing still be an option?**
4. **Based on the cause, what is the likelihood that dystocia will recur with future pregnancies?**
5. **If surgery is performed, how will the puppies be affected? How will the dam be affected?**

Diagnosis: Eclampsia (puerperal tetany, lactation tetany)

Eclampsia is caused by low blood calcium in a dog that is actively nursing pups. Small dog breeds and first-time mothers are most commonly affected. Not only does the dam sacrifice some of her own calcium to the pups during fetal development, the nursing process also drains her calcium stores. Inadequate or unbalanced nutrition and calcium supplementation during pregnancy contribute to the development of eclampsia.

Symptoms include: incoordination; agitation; muscle tremors; a rigid appearance; and even seizures. Diagnosis is based on history, characteristic symptoms, and low blood calcium. Low blood sugar commonly accompanies eclampsia. Treatment consists of injectable calcium administration (and glucose if indicated) followed by oral calcium and temporary discontinuation of nursing. If nursing is to be permanently discontinued a drug called *cabergoline* can be used to help slow milk production.

Questions to Ask Your Vet

1. **Why did my dog develop eclampsia?**
2. **Does nursing need to be permanently discontinued?**
3. **How and what should we feed the puppies? How can we be sure they are receiving adequate nutrition?**
4. **How long should supplemental calcium be given to the dam?**
5. **Should the dam's diet be changed?**
6. **How can eclampsia be prevented in the future?**

Diagnosis: Mastitis

Mastitis refers to inflammation, usually with infection, in the mammary gland(s). It tends to occur in dogs that are nursing pups or experiencing *pseudopregnancy* (see p. 350). The most likely causes are trauma or an infection that travels up the gland from the environment. The affected gland(s) is typically firm, warm, swollen, and painful (pus or discolored milk may be expressed). Dogs with mastitis may be lethargic and lose their appetite and interest in their puppies.

Antibiotics are the mainstay of therapy; antibiotic selection depends on whether or not pups are still nursing. Unless they are close to weaning age, ongoing nursing is recommended. If the puppies are ready to be weaned a medication called *cabergoline* may be given to help stop milk production. Most dogs with mastitis respond to medical therapy alone. On occasion, surgery is needed to drain an abscessed mammary gland. Prognosis is good unless the infection manages to spread to the bloodstream.

Questions to Ask Your Vet

1. **Is more than one gland affected?**
2. **Should the puppies continue to nurse?**
3. **Will there still be enough for all the pups without milk from the affected gland(s)?**

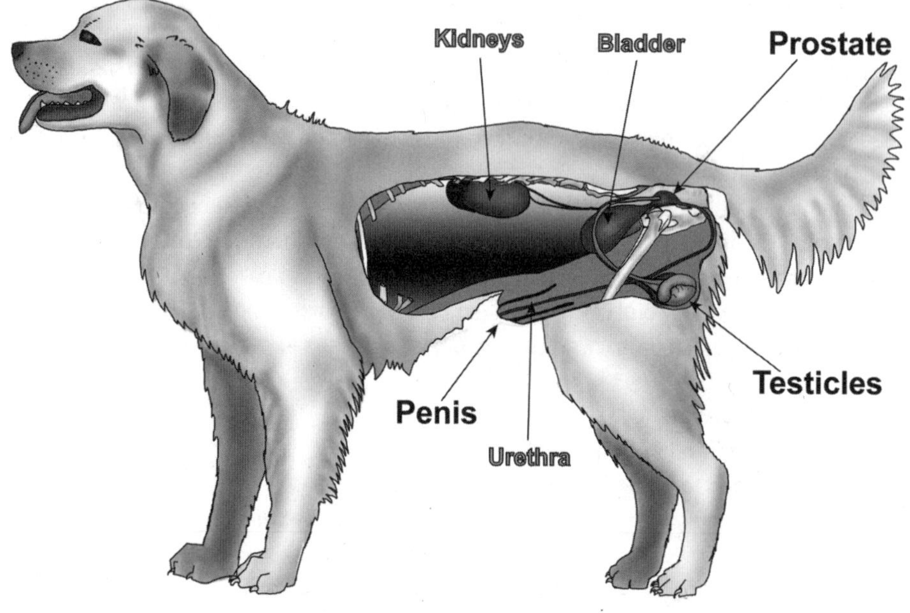

MALE REPRODUCTIVE SYSTEM

❹ **What should I do if my dog no longer wants to be with her puppies?**
❺ **When can I expect her milk to dry up after she stops nursing?**

❊ Don't Forget! Check p. 283 for Universal Questions, and pp. 148 and 152 for questions when dealing with anesthesia and surgery.

Diagnosis: Prostate disease (benign prostatic hyperplasia, bacterial prostatitis, prostatic abscess, prostatic cyst)

The prostate gland surrounds the urethra as it leaves the urinary bladder. It secretes a fluid that enhances sperm fertility. As in men, over time, testosterone causes the prostate gland to enlarge, a process called *benign prostatic hyperplasia (BPH)*. Other diseases of the prostate gland include: infection (*bacterial prostatitis*); cancer; cysts; and abscesses. Depending on the type and severity of the disease, symptoms may include: straining to urinate; straining to pass a bowel movement; hind end discomfort; and discharge (sometimes bloody) from the penis.

Here's some good news! Neutering prevents *all* prostate disease other than cancer, which although rare, occurs equally in both neutered and unneutered dogs. Prostate disease is diagnosed by physi-

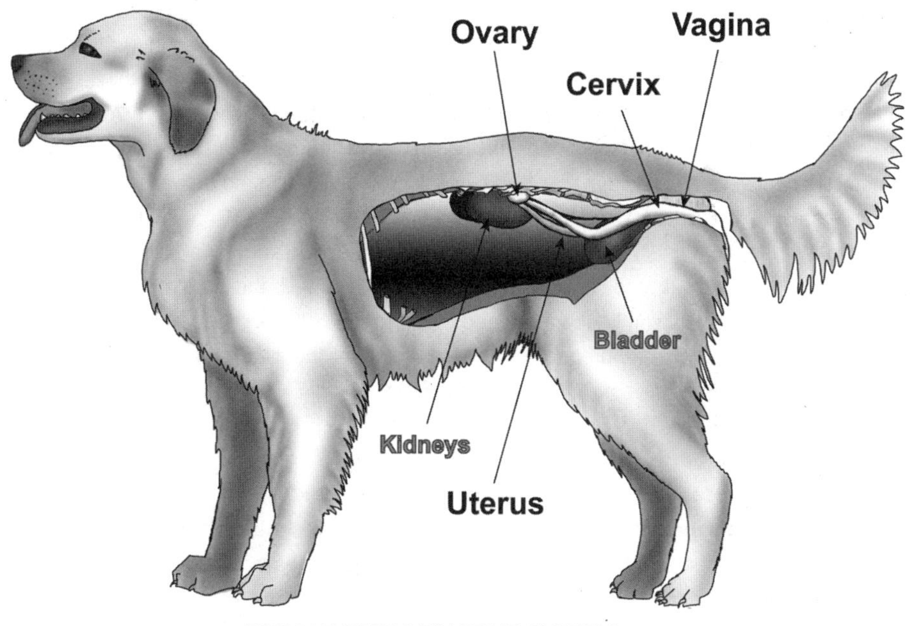

FEMALE REPRODUCTIVE SYSTEM

cal examination findings, imaging studies (ultrasound, X-rays, CT or MRI scans), and sampling of prostatic fluid or tissue.

Specific treatment depends on the nature of the prostate disease, but neutering is universally recommended (for valuable breeding dogs, medical treatment alone can be tried). Antibiotics are used when infection is present; however, only a few antibiotics adequately penetrate into the prostate gland. Prostatic cysts and abscesses often require surgical drainage. The need for pain medication varies with the type of prostate disease. Unfortunately, with prostate cancer, treatment is palliative (aimed at providing patient comfort) rather than curative. The prognosis for noncancerous diseases of the prostate is typically good with appropriate management.

Questions to Ask Your Vet

1. **What type of prostate disease does my dog have?**
2. **Does he need surgery?**
3. **Do you think he is experiencing pain?**
4. **Should neutering be done right away or after treatment with medical therapy (antibiotics)?**

Diagnosis: Pseudopregnancy

A dog experiencing *pseudopregnancy* acts and feels as if she is pregnant even though she is not. It is caused by a hormone called *prolactin*. Although all dogs make prolactin, not all exhibit symptoms of pseudopregnancy, which occur six to 12 weeks following active heat and can include: mammary gland enlargement; excessive licking of the glands; and even milk production. Affected dogs may exhibit various maternal behaviors such as: nesting; being overly protective of stuffed animals or other toys; anxiety; and aggressive behavior toward housemates.

Treatment is usually unnecessary as symptoms typically resolve spontaneously within two to three weeks following onset. In extreme cases, treatment options include: behavioral modification; use of an Elizabethan collar to prevent self-licking; and medication to reduce milk production. Dogs that are predisposed to the effects of pseudopregnancy are likely to suffer the syndrome throughout their reproductive years. Spaying is curative.

Questions to Ask Your Vet

1. **Is my dog a candidate for medication to decrease milk production?**
2. **Do you have any suggestions for modifying my dog's undesirable behaviors?**
3. **Is it likely that subsequent bouts of pseudopregnancy will be as severe?**
4. **If I want to have my dog spayed, when should it be done?**

Diagnosis: Pyometra

Pyometra literally means "pus in the uterus." This condition typically occurs within two to six weeks following a heat cycle. Under the influence of *progesterone*, the uterine lining becomes thickened and produces fluid material, uterine muscle activity diminishes, and normal resistance to infection within the uterus is suppressed. All these factors predispose the uterus to bacterial infection. Dogs with *open* pyometra typically have vaginal discharge because the cervix is open. Vaginal discharge is absent in dogs with *closed* pyometra. Other pyometra symptoms include: lethargy; decreased appetite; fever; vomiting; and increased thirst. Potential complications are spread of infection and subsequent organ failure.

Pyometra is readily diagnosed via abdominal ultrasound in conjunction with: compatible history (the dog was recently in heat); physical examination findings; an elevated white blood cell count; and X-rays. The treatment of choice is removal of the uterus and ovaries (*ovariohysterectomy* or spay procedure) in conjunction with antibiotics. Pyometra can also be treated nonsurgically with antibiotics and medication to help the uterus expel the infection, but there are risks involved and the dog must be bred on her next heat cycle to avoid recurrence of pyometra. Medical therapy is considered only if the dog is highly valued as a breeder and has an open pyometra.

Questions to Ask Your Vet

1. **Does my dog have an open or closed pyometra?**

❷ Does she appear to have any significant complications from her pyometra?

❸ If we treat her medically, what is the likelihood of curing the pyometra?

❹ If we treat the pyometra medically, will she likely be able to become pregnant and carry a litter to term?

Respiratory Tract Diseases

Introduction to the Respiratory System

The respiratory tract consists of the *nose, throat, trachea* (windpipe), and *lungs*. The *larynx* (containing the voice box) is located in the upper portion of the trachea. The *diaphragm* is a thin sheet of muscle separating the chest and abdominal cavities. This muscle—along with the intercostal muscles located between the ribs—helps drive inhalation and exhalation. An easy way to image the respiratory tract is by thinking of it as if it were a tree. During inspiration, air flows through the trunk (trachea), out into the branches of the tree (bronchi) and finally into the leaves (alveoli) where oxygen is delivered to the red blood cells. Simultaneously, carbon dioxide is retrieved from the blood and removed from the body during exhalation.

> QUICK REFERENCE
> **Which Specialist Is Right for My Dog?**
>
> When help is needed with challenging respiratory tract disease cases, most family veterinarians seek assistance from specialists in internal medicine or surgery.

Diagnosis: Brachycephalic airway syndrome (hypoplastic trachea, elongated soft palate, everted laryngeal saccules, stenotic nares)

Brachycephalic means "short-headed" and many dogs have been bred for this short-nosed (what I call "smoosh-faced") appearance—examples include Pugs, Bulldogs, Pekingese, Shih Tzus, Boston Terriers. Brachycephalic dogs often come equipped with: abnormally tiny nostrils *(stenotic nares)*; narrowed windpipes *(hypoplastic tracheas)*; a soft palate that is too long *(elongated soft palate)*; and an abnormal larynx (*everted laryngeal saccules*). All of these abnormalities can result in noisy breathing; increased respiratory rate and effort; and evidence of inadequate oxygenation, such as collapse (inability to rise, fainting), and bluish or purplish gum and tongue color. Breathing difficulties tend to be most apparent during significant exertion, hot weather, and recovery from anesthesia. Obesity exacerbates the upper airway symptoms.

Diagnosis is made via chest X-rays and visual inspection of the nostrils, soft palate, and larynx. Dogs with mild symptoms can be managed with a combination of weight loss, restricted activity, and by avoiding hot, humid, weather—best accomplished by keeping the dog in a cool indoor environment (air-conditioned if need be) during the heat of the day. Surgery can improve some of the structural abnormalities. The prognosis varies with the severity of the airway malformations, obesity, and activity level.

Questions to Ask Your Vet

1. **Which features of brachycephalic airway syndrome does my dog have?**
2. **Are the symptoms severe enough to warrant surgery?**
3. **What can I do at home to forestall a breathing crisis?**
4. **Does my dog need to lose weight?**
5. **What should I do if he is having a respiratory crisis (labored breathing, discolored gums)?**

Diagnosis: Bronchitis (tracheobronchitis, kennel cough)

Bronchitis is inflammation of the air passages (bronchi) in the lungs. Acute bronchitis is often associated with an infection (bacterial, viral, fungal). When the trachea (windpipe) is also inflamed, it is called *tracheobronchitis,* often referred to as "kennel cough." Vaccines are available for some of the agents capable of causing infectious tracheobronchitis. *Chronic* bronchitis is usually a noninfectious inflammatory process caused by allergies, inhaled irritants, and/or structural changes.

Coughing (either dry or productive) is the major symptom associated with bronchitis. Other symptoms are fever, lethargy, decreased appetite, and increased respiratory rate. Kennel cough is diagnosed based on recent exposure to other dogs followed by an abrupt onset of coughing. When the history is not clear-cut or the cough has become chronic, tests are necessary to make a diagnosis. Diagnostic evaluation may include X-rays, blood tests, and airway examination (*bronchoscopy*) with collection of samples.

Therapy for kennel cough includes antibiotics, cough suppressants, and rest. The prognosis is excellent. Treatment of chronic bronchitis includes medication to reduce inflammation, open the airways, and suppress the cough. Antibiotics are used when secondary bacterial infection is present. Weight loss is recommended for obese dogs. The prognosis for bronchitis varies from complete cure to lifelong management.

Questions to Ask Your Vet

1. **Does my dog have an airway infection? If so, is he contagious?**
2. **Should I restrict his activity?**

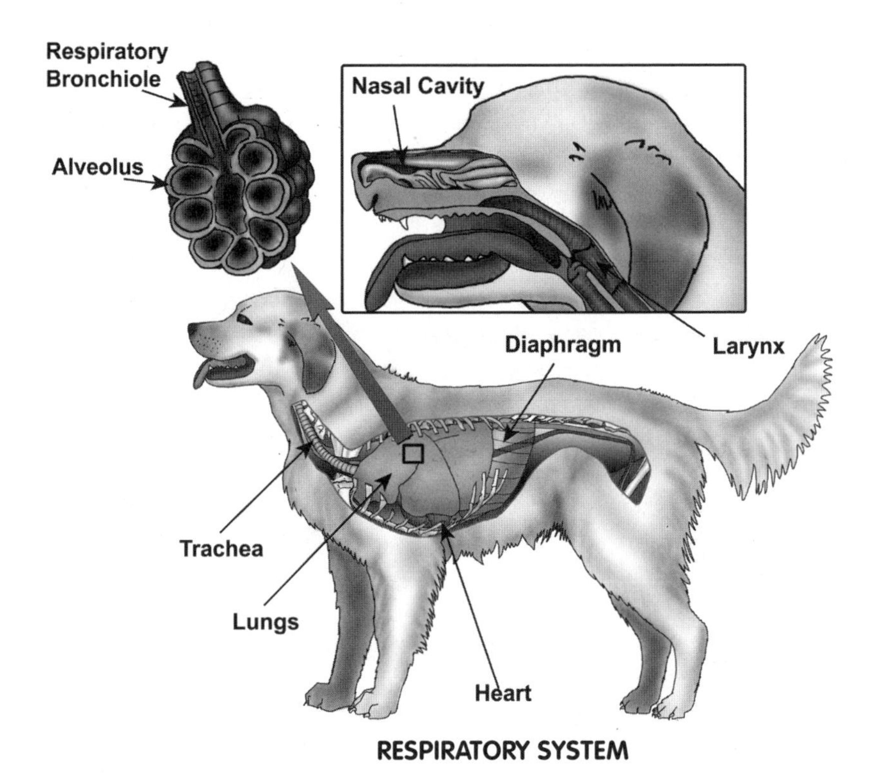

RESPIRATORY SYSTEM

3. **Does he need to lose weight?**
4. **What can I do to minimize airway irritants at home and in the yard?**
5. **Should my dog be vaccinated in the future against infectious tracheobronchitis?**

Diagnosis: Laryngeal paralysis

The larynx (voice box) has two cartilage flaps that work like saloon doors—opening wide during breathing and closing during swallowing or vomiting. With *laryngeal paralysis*, one or both of the laryngeal cartilages does not open normally, creating turbulent, restricted airflow and noisy breathing—especially when the dog is panting. Symptoms worsen during hot weather and exertion as the cartilages become swollen, and can result in a respiratory crisis with labored breathing and cyanotic (blue or purple) coloration of the tongue and gums. Dogs with laryngeal paralysis have an increased risk of *heatstroke* and *aspiration pneumonia* (material intended to be swallowed or vomited is accidentally inhaled into the lungs).

Laryngeal paralysis is caused by an abnormality of the nerves to the larynx; an underlying cause is rarely identified. It occurs most commonly in middle-aged and older, large breed dogs (Labradors top the list). Observing diminished or absent laryngeal motion in the lightly anesthetized dog confirms the diagnosis. Severe breathing difficulties caused by laryngeal paralysis require emergency treatment with oxygen and anti-inflammatory medication. The long-term treatment of choice involves a surgical "tieback" of one laryngeal cartilage. Although surgery increases the risk of aspiration pneumonia, this is outweighed by the benefit of improved airway flow, activity, and comfort.

Questions to Ask Your Vet

1. **Is the laryngeal paralysis interfering with my dog's ability to get enough oxygen?**
2. **Are the symptoms severe enough to warrant surgery?**
3. **Is there any evidence of aspiration pneumonia?**
4. **What precautions can I take at home to prevent my dog from going into a respiratory crisis?**
5. **What should I do when my dog goes into respiratory crisis (labored breathing, discolored gums)?**

❊ **Don't Forget!** Check p. 283 for Universal Questions, and pp. 148 and 152 for questions when dealing with anesthesia and surgery.

Diagnosis: Nasal disease

The nasal passageways extend from the nostrils to the back of the throat. They are connected to the sinuses (air-filled cavities), the largest of which are the frontal sinuses that sit just under the skull on the top of the head. Nasal disease can be caused by: foreign bodies (plant awns called "foxtails" are notorious offenders in the western US); allergies; viral and fungal infections (*aspergillus* is the most common fungal organism); dental disease; nasal mites; cancer; trauma; and a noninfectious inflammatory process known as *lymphoplasmacytic rhinitis*, which is of undetermined cause. Although bacterial infections may occur, they are secondary to another underlying disease process. Disease can originate in a sinus and spread to the nose, or vice versa.

Symptoms of nasal disease include: sneezing; nasal discharge; facial swelling or pain; congested breathing; bad breath; and itchiness around the nose. Diagnostic evaluation may include imaging of the nasal passageways and frontal sinuses (CT or MRI scan, X-rays) and *rhinoscopy* (direct visualization of the nasal cavity). Note: general anesthesia is a must for these procedures in order to keep the dog motionless. Samples from the nasal passageway and sometimes the sinuses are obtained for microscopic evaluation and culture. The frontal sinuses are sampled and flushed (irrigated with a salt solution) if imaging studies demonstrate their involvement. The prognosis and recommended treatment are highly variable depending on the underlying cause and extent of the disease.

Questions to Ask Your Vet

1. **Do we know what is causing my dog's nasal disease? If so, how can it be treated? If not, what can be done to determine what it is?**

❷ **Does the disease involve one or both nasal passageways?**
❸ **Is there frontal sinus involvement?**
❹ **Do you think my dog is experiencing pain?**
❺ **Is it likely we can cure this or are we looking at a long-term management issue?**

Diagnosis: Pleural disease (pleural effusion, pneumothorax)

The pleural cavity is the normal space between the lungs and the inside of the chest wall. Several diseases cause fluid (*pleural effusion*) or air (*pneumothorax*) to accumulate in this space, interfering with the lungs ability to expand, resulting in increased respiratory effort. Causes of fluid accumulation within the lungs include: trauma; heart failure; infection (pyothorax); decreased blood protein levels; leakage of lymph fluid (chylothorax); and cancer. Pneumothorax usually occurs with air leaking from the lungs caused by trauma, pneumonia, a migrating foreign body, or cancer.

Pleural effusion and pneumothorax can readily be diagnosed with chest X-rays. Determining the underlying cause often requires further testing, such as ultrasound, CT or MRI scan, and analysis and culture of the pleural fluid.

Treatment and prognosis depend on the underlying cause. If breathing is significantly impaired, removal of the fluid or air from the pleural cavity is recommended. Sometimes a temporary drain may be placed to allow ongoing removal of fluid or air.

Questions for Your Vet

❶ **Does my dog have pleural effusion or pneumothorax?**
❷ **Has the underlying cause of the pleural fluid or air been diagnosed? If so, how can it be treated? If not, how can the cause be determined?**
❸ **If we suspect trauma, is there evidence of damage to other organs?**
❹ **Does the fluid or air need to be removed to help my dog's breathing? Once removed is it likely to recur?**
❺ **Is my dog in need of supplemental oxygen therapy?**

Diagnosis: Pneumonia

Pneumonia is an infection in the lungs. *Bacterial* infections are the most common cause of pneumonia, but *fungal* and *viral* pneumonias also occur. It is important to look for an underlying cause such as: an inhaled foreign body; the spread of infection from another site in the body; aspiration of material that has been vomited or regurgitated; or suppression of the normal immune system by medication or another disease. With rare exceptions, pneumonia is not contagious from dog to dog—or from dog to human.

Common symptoms of pneumonia include: fever; lethargy; decreased appetite; coughing; and labored breathing. Pneumonia typically causes an increased white blood cell count and characteristic lung changes seen on X-rays. Antibiotics are the mainstay of therapy for bacterial pneumonia; antifungal medications for fungal pneumonia. Hospitalization to give intravenous fluids and antibiotics is recommended for severely affected dogs. Supplemental oxygen therapy may also be indicated. Progress is determined based on follow-up chest X-rays along with resolution of the dog's symptoms. If the underlying cause of the pneumonia is curable, the prognosis for complete recovery from the pneumonia is good.

Questions to Ask Your Vet

1. **Do we know what caused my dog's pneumonia? If so, what can be done to treat it? If not, what can be done to determine the cause?**
2. **Does he need supplemental oxygen?**
3. **If he needs to be hospitalized, how will he be monitored overnight? What treatments will he receive?**
4. **How soon will more X-rays be taken?**
5. **Should I limit his activity level at home? If so, for how long?**

Diagnosis: Tracheal collapse

The trachea (windpipe) carries oxygen from the throat to the lungs. It is made up of smooth, round rings of cartilage. Some small breed dogs have a weakness of the cartilage that causes a section of the trachea to collapse. This results in a "goose-honk" cough, readily induced by activity, exercise, or even simply pulling on the dog's collar. The cough can be persistent, keeping both dog and human awake at night. At its worst, tracheal collapse can cause: labored breathing; discolored (purple or blue) gum and tongue color; and syncopal (fainting) or collapsing episodes. The cause of this disease is uncertain, but genetics and obesity likely play a role.

X-rays are commonly used to confirm diagnosis. Since tracheal collapse is often associated with a particular phase of respiration, it can be missed with plain X-rays. For this reason, fluoroscopy (an x-ray that is in a movie format) or tracheoscopy (a telescope device that is placed into the trachea) can be used.

Most dogs with tracheal collapse respond well to restriction of activity, cough suppressants, and anti-inflammatory medication. Sometimes, simply breaking the coughing cycle with medication provides extended relief. Weight reduction may also be helpful. Dogs with severe disease require hospitalization for oxygen therapy (either within an oxygen cage or via oxygen-carrying tubes placed within the nostrils). Surgery to insert synthetic rings within the trachea can be successful but has a high complication rate and is considered only as a last resort.

Questions to Ask Your Vet

1. **Is the tracheal collapse affecting my dog's ability to get enough oxygen?**
2. **What can I do to minimize tracheal irritation/stimulation?**
3. **Should I modify my dog's activity level?**

❹ **Does he need to lose weight?**
❺ **Should I expect his cough to go away completely? How can we get a good night's sleep?**
❻ **What should I do if he is having a respiratory crisis (labored breathing, discolored gums)?**

✻ **Don't Forget!** Check p. 283 for Universal Questions, and pp. 148 and 152 for questions when dealing with anesthesia and surgery.

Skin and Ear Diseases

Introduction to the Skin and Ears

Believe it or not, the skin is considered a vital organ in the body. The ear canals are an extension of the skin. It makes sense, therefore, that many dogs with ear disease have skin disease, and vice versa. The ear consists of the *external ear canal*, the *middle ear canal* (where the hearing apparatus is located), and the *inner ear*, which receives auditory signals and contains the *vestibular apparatus* (responsible for maintaining normal balance).

QUICK REFERENCE

Which Specialist Is Right for My Dog?

Board certified veterinary dermatologists are the specialists best equipped to tackle skin and ear problems that are difficult to diagnose or refractory to therapy. If a dermatologist is not available in your area, referral may be made to a specialist in internal medicine.

Diagnosis: Acral lick dermatitis (lick granuloma)

Acral lick dermatitis is a chronic skin sore caused by excessive licking. The lower leg is the usual site and large breed dogs are the typical culprits. There is great debate about whether the problem is behavioral (similar to thumb-sucking) or medical (underlying arthritis, trauma, allergy). The diagnosis is made based upon breed, history, and visual inspection.

Lick granulomas require long-term therapy and some trial and error is usually necessary to come up with the best approach. Cure may be unrealistic because most dogs remain intent on licking, happy to start in on a new leg as soon as the old sore is healing (they seemingly just always need a site to lick). Elizabethan collars, topical anti-inflammatory medication, and taste deterrents are common treatment strategies. Antibiotics are used for secondary infections. Behavioral therapy runs the gamut from anti-anxiety medication to providing increased stimulation.

Questions to Ask Your Vet

1. **Do you know what is causing my dog to lick his leg?**
2. **Is the sore infected?**
3. **In your experience, which treatment strategy has been most successful?**
4. **Is it harmful to let my dog continue to lick as long as the skin isn't getting worse?**

Diagnosis: Acute moist dermatitis (hot spot)

Acute moist dermatitis results from self-inflicted trauma (scratching, biting, licking) with secondary infection. One minute your dog's skin and coat look perfect; the next, you discover a red, warm, oozing patch of skin. Infections, fleas, allergies, and skin trauma can all result in "hot spots." Clipping the fur to allow air to the wound, prevention of ongoing self-trauma, and medication to control inflammation and infection are the mainstays of therapy.

Questions to Ask Your Vet

1. **Do you know what caused my dog to create his hot spot?**
2. **How can I prevent it from recurring?**
3. **How long will I need to take measures to prevent my dog from licking and scratching at the site?**

Diagnosis: Atopy (allergic dermatitis)

Atopy is itchy skin caused by allergies to things that are either inhaled (pollens, mold, house mites) or absorbed through the skin. This disease is thought to be inherited and certain breeds (Terriers in particular) are commonly affected. Dogs with atopy commonly have a seasonal itch primarily affecting the feet and face. Over time, however, the itchy season becomes longer and more of the body becomes affected. Secondary bacterial and yeast infections are common, as are ear infections.

The diagnosis of atopy is made by ruling out other causes of itchy skin. Intradermal skin testing and blood testing can be done to help identify which substances might be causing the allergic response. Atopy cannot be cured and rarely can the substance(s) inducing the allergy be eliminated from the environment. Treatment options include: medicated shampoos; antihistamines; omega-3 fatty acids; and anti-inflammatory medications. Treatment with cortisone may be unavoidable, but because of negative side effects, its use should be limited to the least amount necessary for the shortest time period. *Cyclosporine* is a medication that has a great track record treating atopy and has helped minimize the need for cortisone. *Immunotherapy* (allergy injections) to desensitize the dog to his allergies is beneficial in approximately 70 percent of dogs with atopy.

Questions to Ask Your Vet

1. **Does my dog have any skin diseases in addition to his atopy? If so, how should they be treated?**
2. **Does he also have flea or food allergies?**
3. **Is intradermal skin or blood testing warranted to try to figure out what my dog is allergic to? If so, which type of testing is preferred?**
4. **How aggressive do we need to be with therapy?**

Diagnosis: Demodecosis (demodectic mange)

Demodecosis (also known as *demodectic mange*) describes a skin disease caused by microscopic mites called *Demodex canis*. These mites are normal residents of a dog's hair follicles and are quite harmless unless his immune system happens to let down its guard, allowing the mites to undergo a population explosion. Not to worry, demodecosis is not transmissible to people nor are the mites transmissible from dog to dog other than during the first few days of a puppy's life.

Demodecosis symptoms include: crusting of the skin; hair loss; and itchiness. Young dogs are most commonly affected, presumably due to an "immature" immune system. Most outgrow the disease by one year of age. Nonetheless, it is recommended that affected dogs not be bred. When demodecosis occurs in an older dog, it is invariably the result of medication (chemotherapy, cortisone) or an underlying disease process (hormonal imbalance, infection, tumor) that is suppressing the immune system. Pregnancy and a heat cycle can also trigger demodecosis.

Demodectic mange is readily diagnosed by performing a simple skin scraping. The affected skin is gently abraded and the surface material gathered and examined for the presence of mites under the microscope. A skin biopsy can also be diagnostic.

A young dog with *localized* demodecosis (affecting only a few sites) might require nothing more than benign neglect. Topical medications are sometimes used. When large areas of the body are affected (*generalized* demodecosis), oral medications and/or topical dips are recommended. Therapy for dogs over a year of age is aimed at treating the underlying cause of immunosuppression. Bacterial infections of the skin commonly accompany demodecosis and, for these dogs, antibiotic therapy provides great benefit. For most cases of localized demodecosis in young dogs, the prognosis is good.

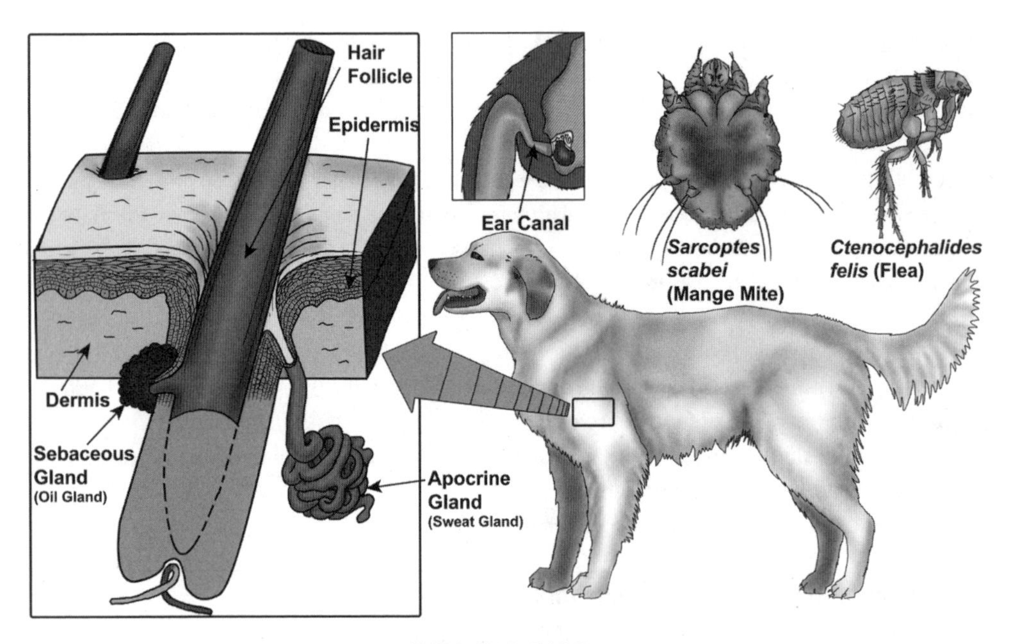

SKIN AND EARS

Otherwise, the prognosis for demodecosis is highly variable.

Questions to Ask Your Vet

1. **Does my dog have localized or generalized demodecosis?**
2. **If the infection is localized, is treatment necessary?**
3. **Do we know what has caused immune system suppression in my older dog with demodecosis? If so, what can be done to treat it? If not, how can we figure it out?**
4. **Does he have a secondary bacterial infection?**
5. **How will we know when we can stop the medication?**

Diagnosis: Discoid lupus erythematosus (nasal solar dermatitis, "Collie nose")

Discoid lupus erythematosus (DLE) is a crusty, inflamed skin condition that affects the nose and adjacent unhaired portion on the bridge of the nose. If the DLE is longstanding, the nose loses its pigmentation and the surface may become eroded. Occasionally, the lips, ears, and skin around the eyes are affected. DLE affects primarily long-nosed breeds of dogs (Collies, German Shepherds). DLE is an immune mediated skin disease, meaning that the immune system attacks the body's own tissues. Sun exposure is thought to play a role in initiating this immune system "bad behavior."

Diagnosis is based on the characteristic appearance of the nose. A biopsy might be recommended. In addition to sun avoidance, treatment is comprised of topical and oral medications to suppress the overactive immune system. If a biopsy is not performed, antibiotics are used to treat for a disease that can mimic DLE called *mucocutaneous pyoderma*. The prognosis is good, but long-term treatment is usually required.

Questions to Ask Your Vet

1. **Do you recommend a biopsy? If so, why?**
2. **Is my dog likely to develop other immune mediated diseases?**
3. **My dog loves the sun—how do I keep him happy while dealing with this disease?**

✻ **Don't Forget!** Check p. 283 for Universal Questions, and pp. 148 and 152 for questions when dealing with anesthesia and surgery.

Diagnosis: Flea allergy dermatitis (flea bite allergy)

We all know that fleas cause dogs to itch, but some dogs take it to an extreme. Dogs who are allergic to flea saliva suffer long after the flea has gone, and it only takes one lousy flea! Your veterinarian doesn't need to see the actual flea because she will recognize the characteristic allergy pattern of skin inflammation, including redness, crusting, and evidence of self-trauma at the base of the tail. The dermatitis may extend to the back legs and underside of the belly. Intradermal skin and blood testing are available to help confirm a diagnosis of flea bite allergy, but these tests are not 100 percent accurate.

Treatment consists of flea control (including all of the household's dogs and cats, as well as their home and yard environment) and anti-inflammatory medications such as cortisone and antihistamines to "cool off" the angry skin. Antibiotics will be prescribed if a secondary bacterial skin infection is present. Flea prevention is critical as there is no cure for the underlying allergy. The ultimate treatment is a move to Colorado as fleas do not fare well in high-altitude regions!

Questions to Ask Your Vet

1. **Which products do you recommend for flea control on my dog(s)?**
2. **Which products do you recommend for flea control on my cat(s)?**
3. **Which products do you recommend for flea control in the home environment? Which for the yard environment?**
4. **Is there a way to manage my dog's flea allergy using minimal or no cortisone?**
5. **Should I avoid taking my dog to places where other dogs might expose him to fleas?**

Diagnosis: Food allergy (food hypersensitivity, dermatologic adverse reaction to food)

When dogs are allergic to something in their diet, they commonly develop itchy skin. There is no particular pattern of itch—any body part can be affected. Many food-allergic dogs develop concurrent allergies to fleas and inhaled allergens. The most common foods implicated include beef, chicken, dairy products, eggs, wheat, corn, and soy. Unlike other allergies that are more seasonal, food allergies tend to occur year round.

The diagnosis of food allergy is confirmed by ruling out other skin diseases and by observing a clear and dramatic response to an elimination food trial (neither blood testing nor intradermal skin testing is accurate when it comes to food allergies). Some vets believe that true confirmation requires a dietary challenge that involves reintroducing the original food to see if itchy skin recurs.

A number of commercially prepared "novel protein diets" (containing a type of protein that the dog has never before been exposed to) are available with main ingredients such as white fish, rabbit, duck, venison, and even kangaroo! The theory is that if your dog hasn't eaten an ingredient in the past, he likely won't be allergic to it. One of these novel protein foods or a homemade diet limited to just a couple of ingredients may be recommended for a challenge food trial. Ideally, this food trial is conducted without concurrent anti-inflammatory medications such as cortisone or antihistamines. Short-term therapy involves anti-inflammatory medication and treatment of concurrent skin and ear infections. Long-term therapy is strict adherence to a suitable diet, recognizing that new food allergies may develop with time. The prognosis for dogs with food allergies is excellent when the offending foods are avoided.

Questions to Ask Your Vet

1. **How long will my dog need to eat the new food before we know if it's working?**
2. **Can I continue to give him his usual chew toys, treats, vitamins, and heartworm preventative? If not, what can be substituted?**
3. **If I want to feed a homemade diet, how can I be sure it is nutritionally complete?**
4. **Is it okay to feed my other dog(s) the same diet?**

Diagnosis: Hygroma (elbow pressure sore)

A *hygroma* is a fluid-filled pocket that arises under the skin on the outside surface of the elbow. This is a major pressure point when dogs lie down and, the bigger the dog, the greater the pressure. It makes sense that most dogs with hygromas are large or giant breeds. Not surprisingly, obese dogs and older dogs that spend a good deal of their time lying down are more susceptible.

Diagnosis is based on visual appearance. Treatment consists of draining the fluid from the hygroma (an indwelling drain might be placed) and the use of soft, padded bedding in the hopes of allowing the pressure sore to heal. Unfortunately, padded elbow wraps do not work well. Antibiotics will be prescribed if the hygroma fluid appears infected. Surgical removal of the hygroma is truly a last resort, because proper wound healing may be impossible.

Questions to Ask Your Vet

1. **What is the ideal type of padding for my dog?**
2. **Does he have an orthopedic issue that is making him lie down more than normal? If so, what can be done to make him more comfortable?**
3. **Is the hygroma infected?**
4. **Is one or are both elbows affected?**
5. **My dog prefers cold hard surfaces. Do you have any pointers to encourage my dog to lie on a padded surface?**
6. **What should I do if it appears that the hygroma is beginning to recur?**

Diagnosis: Otitis externa (ear infection)

Otitis externa is inflammation of the outer ear canal, a tubular structure that contains a vertical portion, connected by a 90-degree turn to a horizontal portion that ends at the ear drum. The ear canals are a direct extension of the skin. Anything that creates skin inflammation is capable of creating otitis. Allergies, parasites, bacteria, yeast, foreign bodies, tumors, hormonal imbalances, and retained moisture in the ears following bathing or swimming are the most common causes.

Symptoms include: head-shaking; pawing at the ears; discharge; redness; and odor. Diagnosis is made by visual inspection of the ear canal using an otoscope (this can be uncomfortable for dogs with otitis) and by finding bacteria, yeast organisms, or white cells under the microscope when inspecting discharge obtained from the ear canal. Treatment is based on the underlying cause of the otitis and whether or not infection is present. Topical medications are sufficient for dogs with mild otitis. Oral drugs may be recommended when the degree of inflammation or infection is severe.

Some dogs are predisposed to getting ear infections over and over again. Lifelong preventive and maintenance therapy may be needed. Chronic otitis can lead to permanent scarring and narrowing of the ear canal, in which case a surgical procedure to open up or actually remove the ear canal may be recommended.

Questions to Ask Your Vet

1. **What is the cause of my dog's otitis?**
2. **Are both ears affected?**
3. **Should I treat his ears with medication after bathing or swimming?**
4. **Should I clean his ears on a regular basis?**
5. **Should the groomer do anything special?**
6. **Should the hair be removed from my dog's ears?**

Diagnosis: Pemphigus

Pemphigus refers to a group of autoimmune skin diseases in which the immune system is triggered to attack the body's own skin tissues. Pemphigus does have some breed predilection, but also seems to be exacerbated by sunlight. Some cases begin as reactions to medication or as a result of chronic skin diseases such as allergic dermatitis. Typical symptoms include painful crusting and scaling that begin on the face and ears before spreading to other sites. The footpads are commonly affected and secondary bacterial infections frequently occur. The diagnosis is made via skin biopsy.

Medications that suppress the immune system are the mainstay of therapy. Antibiotics and pain medication are used as needed and avoidance of sun exposure is typically recommended. Response to treatment is highly variable and long-term treatment usual. Unfortunately, because of the pain associated with pemphigus, dogs that fail to respond to medical therapy are commonly euthanized.

Questions to Ask Your Vet

1. **What form of pemphigus does my dog have? What is the usual course of this form?**
2. **Is there a secondary skin infection?**
3. **What is the best way to gauge whether or not pain is being adequately managed?**
4. **Must sunlight be completely avoided?**
5. **How long will it take before we begin to see improvement?**
6. **Is it likely that my dog will develop other immune mediated diseases?**

Diagnosis: Pododermatitis (interdigital dermatitis, interdigital pyoderma)

Pododermatitis refers to inflammation of the skin affecting the paws. It can arise from contact irritation, allergies, infections, foreign bodies, and immune mediated disease (the immune system attacks the body's own tissues). Symptoms can include: excessive licking and chewing at the feet; hair loss; inflammation; tenderness; firm nodules; and open sores. The nail beds may be involved and regional lymph nodes can be enlarged.

Diagnostic testing is aimed at identifying an underlying skin disorder. Treatment involves correcting or avoiding any underlying conditions and may include: medicated shampoos; oral antibiotics; removal of a foreign body; and treatment for mange or allergic skin disease. Prognosis is dependent on the underlying cause. Many cases of pododermatitis require long-term—if not lifelong—therapy.

Questions to Ask Your Vet

1. **What has caused my dog's pododermatitis? If unknown, how can we find out?**
2. **Do you think he is experiencing much discomfort?**
3. **Do I need to restrict his activities or be careful about what his feet are exposed to? Is it okay for him to swim?**
4. **Is it likely that chronic treatment will be necessary?**

❋ **Don't Forget!** Check p. 283 for Universal Questions, and pp. 148 and 152 for questions when dealing with anesthesia and surgery.

Diagnosis: Pyoderma (bacterial dermatitis, impetigo, acne)

Pyoderma is a bacterial infection of the skin. Pyoderma is characterized as *superficial* or *deep*, depending on which layers of the skin are affected. *Acne* is localized to the chin. *Impetigo* refers to the superficial pyoderma that commonly occurs in puppies. German Shepherds are the number-one breed afflicted with pyoderma, but it can occur in any breed, often secondary to another skin disease. *Staphylococcus* bacteria that normally reside on the skin surface cause approximately 90 percent of pyoderma cases. Anything that inflames the skin offers these bacteria the opportunity to propagate and penetrate deeper skin layers. Hormonal imbalances and medication that suppress the immune system commonly cause pyoderma.

Symptoms vary from surface pustules (pimples), to scaling, crusting open sores and deep draining tracts. Some dogs manage to develop an allergy to the *Staphylococcus* organisms, resulting in itchy skin and self-trauma. Diagnosis is based on characteristic appearance and microscopic identification of bacteria from the lesion (skin abnormality). Antibiotics (a minimum of three- to four-weeks-worth) and medicated shampoos are the mainstay of treatment. Bacterial culture (identification of the type of bacteria) and antibiotic sensitivity testing is warranted in dogs with deep pyoderma. Some dogs are given subcutaneous (under the skin) injections of an immunomodulator intended to desensitize them to the bacterial allergy. Prognosis for superficial pyoderma is excellent. Prognosis for deep pyoderma is dependent on underlying cause and response to treatment. Many cases are controlled only with long-term antibiotic therapy.

Questions to Ask Your Vet

1. **Does my dog have another skin disease that has caused him to develop the pyoderma?**
2. **Does he have a hormonal imbalance that has caused him to develop the pyoderma?**
3. **Do you suspect he is allergic to the Staphylococcus?**
4. **Is it likely he will require chronic antibiotic therapy? What other therapies might we consider?**

Diagnosis: Sarcoptic mange (canine scabies, mange)

Sarcoptes scabei is a microscopic mite capable of creating horribly itchy skin. They are highly contagious from dog to dog. Although humans are not the preferred host, we can be infested from contact with an affected dog. Dogs with *sarcoptic mange* itch mostly around the elbows and ears where there may be redness and crusting. Many dogs with sarcoptes have a positive "ear-scratch reflex": when the margin of the ear is rubbed, the dog has a reflexive hind-leg scratching motion.

Skin scrapings are made to diagnose sarcoptic mange. The affected skin is gently abraded and the surface material gathered and examined under the microscope. The presence of the mites or their eggs confirms the diagnosis; however, it takes very few mites to create an itchy dog; one has to have a bit of luck to find mites even in scrapings from multiple sites. If the scrapings are negative but there remains a high level of suspicion, many veterinarians will treat for sarcoptic mange to see if the itch goes away. Treatment consists of oral or injectable *mitecide* medication. Be sure to tell your veterinarian if your dog is part Collie, as this breed has an inherent sensitivity to *ivermetin*, a common treatment for sarcoptic mange. Antibiotics are used when a secondary skin infection is present.

Questions to Ask Your Vet

1. **Were mites and/or eggs seen on the skin scrapings?**
2. **Should my other dogs be tested and/or treated?**
3. **Do the home and yard environments need to be treated?**
4. **How long will therapy continue after my dog has stopped itching?**
5. **How soon can he resume his normal activities that involve other dogs?**
6. **What precautions do family members need to take to avoid getting mites?**

Diagnosis: Seborrhea

Seborrhea is a skin disorder resulting in scaling of the skin, often accompanied by excessive sebum production, resulting in a greasy feel to the skin and coat. Seborrhea commonly occurs as a result of other skin disorders (parasites, allergies, immunological diseases) or hormonal imbalances. Some breeds (Cocker Spaniels easily top the list) are predisposed to seborrhea.

Seborrheic dogs tend to have scaly, unthrifty-appearing, smelly hair coats. Scales found on the skin surface or commingled with the hair are readily apparent. Ear infections are common. The diagnosis of seborrhea is confirmed with a skin biopsy. Treatment includes medicated shampoos and fatty acid supplements. Vitamin A results in a dramatic improvement in some Cocker Spaniels with seborrhea. Zinc has a similar impact on some affected arctic breeds (Siberian Husky, Samoyed, Alaskan Malamute). *Retinoids* (a man-made relative of vitamin A), antibiotics, and a drug called *cyclosporine* are also used in some dogs with seborrhea. Many dogs require lifelong therapy.

Questions to Ask Your Vet

1. **Does my dog have any concurrent skin or health issues that are causing the seborrhea? If so, how can they be treated?**
2. **Are my dog's ears affected? If so, how can they be treated?**
3. **How often can I bathe him? What should I use?**
4. **What nutritional supplements do you recommend?**

Diagnosis: Yeast dermatitis (Malassezia dermatitis)

Malassezia pachydermatis is the fancy name for the yeast organism that can cause intensely itchy skin, hair loss, and *seborrhea*. It is also a common cause of ear infections. Small numbers of Malassezia organisms are normally found on the skin, but problems arise when a yeast population explosion occurs secondary to underlying skin disease or illness. Humid conditions and drugs that suppress the immune system can also cause *yeast dermatitis*. Some dogs even develop an allergy to the yeast organism. Dogs with Malassezia dermatitis tend to have itchy skin, a greasy feeling haircoat, and a characteristic unpleasant "yeasty" odor.

The diagnosis is readily made by microscopic evaluation of a smear from the affected skin surface. Mild cases of yeast dermatitis can be treated with medicated shampoos. Oral antifungal medication is reserved for more serious infections. The prognosis is dependent on the ability to detect and treat any underlying disorders.

Questions to Ask Your Vet

1. **Does my dog have any concurrent skin or health issues that are causing the yeast infection? If so, how can they be treated?**
2. **Are his ears affected? If so, how can they be treated?**
3. **Is he a candidate for oral antifungal therapy?**
4. **How will we know when the yeast infection has fully resolved?**

Urinary Tract Diseases

Introduction to the Urinary Tract

The *upper* urinary tract consists of the kidneys and ureters. Urine is formed in the kidneys and then transported down the ureters into the bladder, which serves as a holding reservoir. The bladder and urethra comprise the *lower* urinary tract. When the bladder becomes distended, a signal is sent to the brain, which in turn signals the urethra to relax and allow urine to flow freely. The urinary system is responsible for filtering a variety of toxic and metabolic substances from the bloodstream for removal from the body.

QUICK REFERENCE

Which Specialist Is Right for My Dog?

When help is needed with challenging urinary tract disease cases, family veterinarians seek assistance from specialists in internal medicine or surgery.

Diagnosis: Glomerulonephropathy (glomerular disease, glomerulonephritis, glomerulopathy, protein losing nephropathy)

Inside each kidney are millions of *glomeruli*, microscopic structures that filter and retain protein (albumin) while allowing fluid to pass into the urine. With disease, the glomerular "sieve" becomes leaky, allowing large quantities of protein to be lost into the urine. This protein leakage remains silent, causing no symptoms until significant damage has occurred. Left unchecked, *glomerulonephropathy (GN)* can result in dangerously low blood protein levels, kidney failure, blood clot formation, and high blood pressure. Potential symptoms include: fluid retention; lethargy; decreased appetite; vomiting; and diarrhea. GN is diagnosed via a simple urine test called a *protein to creatinine ratio*.

GN is caused by damage inflicted by the body's own immune system, often secondary to infectious diseases, tumors, and a hormonal imbalance called *Cushing's syndrome* (see p. 298). Identification and treatment of the underlying cause offers the best chance of a cure. A kidney biopsy rarely changes treatment, but may offer insight about the prognosis. Treatment of GN typically includes medication to: control high blood pressure; reduce blood clot formation; preserve kidney function; and decrease protein loss in the urine. The prognosis is highly variable and depends on the underlying disease process.

Questions to Ask Your Vet

1. **Do we know the cause of my dog's glomerular disease? If not, what can be done to try to determine this?**
2. **Does he have high blood pressure?**
3. **Does he have kidney failure?**
4. **Does he have evidence of any other immune system damage?**
5. **Should the protein level in my dog's diet be altered?**

Diagnosis: Kidney failure (renal failure)

The main function of the kidneys is to eliminate waste products from the body. When the kidneys cannot keep up with this task, *kidney failure* results. The kidneys show no sign of failure until at least 75 percent of their total function is compromised (the reason why many species, dogs and people included, can donate a kidney, yet still maintain normal function). Causes of failure include: congenital (from birth) defects; infections; toxins (ingestion of antifreeze, grapes, raisins, lily plants); medication; urinary tract obstruction; immune system disease; cancer; heart failure; elevations in blood calcium; and age-related degeneration.

Kidney failure is diagnosed based on increased *blood urea nitrogen (BUB)* and *creatinine* (substances normally cleared by the kidneys) in conjunction with dilute urine. Determining the cause of the kidney failure relies on: history (especially for toxin exposure); imaging studies (X-ray, ultrasound, CT scan); specialized blood testing; and in some cases, analysis of kidney tissue or fluid samples. Symptoms of kidney failure include: increased thirst; increased urine volume; lethargy; diminished appetite; vomiting; diarrhea; weight loss; and halitosis (bad breath). Hypertension (high blood pressure), anemia, and changes in calcium/phosphorus balance often accompany kidney failure, especially when chronic.

If treated early, acute renal failure can sometimes be reversed. Hospitalization for intravenous fluid therapy is usually recommended, and in extreme cases, kidney dialysis may be considered. Once kidney disease becomes chronic, it is invariably progressive. The goal of treatment is to slow this progression and compensate for the loss of normal kidney function. Treatment includes diet change, medication, and supplemental fluids. Frequent monitoring is important. At the time of publication, canine kidney transplantation has met with limited success.

Questions to Ask Your Vet

1. **Do we know what caused my dog's kidney failure? If not, how can we find out?**
2. **Does my dog have acute or chronic kidney failure? If acute, what is the prognosis for returning to normal kidney function? If chronic, what is the prognosis for slowing the progression of the failure?**
3. **What, if any, complications of kidney failure does my dog have (high blood pressure, anemia, calcium/phosphorus imbalance, protein leakage into the urine, ulcers within the mouth or upper gastrointestinal tract)?**

Diagnosis: Urinary incontinence (hormone responsive incontinence)

Urinary incontinence (urine leakage) most commonly occurs when a dog is sleeping. Sometimes the leakage is continuous. Causes include: neurological diseases; urinary tract infections; and structural abnormalities of the urethra, vagina, and ureters. Excessive thirst can exacerbate urinary incontinence by causing bladder overfilling. Some causes of urine leakage benefit from surgical treatment, others from medical therapy.

Hormone responsive incontinence, the most common form of urinary incontinence, occurs primarily in middle-aged and older spayed female dogs. Caused by a laxity of the urethral muscle (*sphincter*), it is often successfully treated with an estrogen compound, *diethylstilbesterol (DES)*, or a drug called *phenylpropanolamine*. Injections of "bulking agents" around the urethra are also an option for some dogs.

Questions to Ask Your Vet

1. **What is the cause of my dog's urinary incontinence? If we don't know, what can we do to find out?**
2. **Does the type of incontinence my dog has pose an actual health risk?**
3. **Are other symptoms likely to develop?**
4. **Should I withhold water at night to try to decrease urine production?**
5. **Can my dog be fitted for a diaper?**

Diagnosis: Urinary tract infection (bacterial cystitis, pyelonephritis)

Urinary tract infection refers to the presence of bacteria within the normally sterile urinary bladder (*bacterial cystitis*) or kidneys (*pyelonephritis*). Typical symptoms of a bladder infection are: frequent urination; straining to urinate; inappropriate urination (break in housetraining); and changes in normal urine characteristics (color, odor, volume). Kidney infections can be associated with loss of normal kidney function. Potential symptoms include: lethargy; decreased appetite; vomiting; diarrhea; change in thirst; and back pain.

Common causes for urinary tract infections include: infection within other regions of the reproductive and urinary tracts; physical abnormalities (stones, tumors, anatomical defects); some hormonal imbalances; and medications that suppress normal immune system activity (chemotherapy drugs, cortisone). An underlying cause for infection cannot always be found, but when a dog has recurrent urinary tract infections, it is important to investigate. Blood tests and imaging studies (abdominal X-rays, ultrasound) may be recommended.

Infection within the urinary tract is best documented by a urine culture, and the most accurate test results are obtained when urine samples are collected via *cystocentesis*, in which a small needle is inserted directly into the bladder (most dogs react no differently to this than they would to a vaccination). If infection is present, bacterial colonies will grow within 24 to 72 hours and can be tested against various antibiotics to determine the best therapy. A *urinalysis* (microscopic evaluation of the urine) can also document infection but isn't as accurate. Ultrasound of the abdomen is helpful for determining the presence of infection within the kidneys.

Hospitalization for intravenous fluids and antibiotics may be indicated when a kidney infection is present. Dogs with bladder infections are typically treated at home. Antibiotic choice is ideally based on urine culture results. Dogs with recurrent infections require frequent monitoring and may require long-term, low-dose antibiotic therapy (the antibiotic dose is decreased and is given just once a day at bedtime). Follow-up urine cultures are essential because the resolution of symptoms doesn't always correlate with clearance of the infection.

Questions to Ask Your Vet

1. **Was the infection documented by culture or urinalysis?**
2. **Was antibiotic sensitivity testing performed? If so, what were the results?**
3. **Is there evidence of kidney involvement? If so, has it affected my dog's kidney function?**
4. **Do we know what caused his urinary tract infection? If so, how should that be addressed? If not, how could we look for the underlying cause?**
5. **How frequently will we be performing urine cultures?**

* **Don't Forget!** Check p. 283 for Universal Questions, and pp. 148 and 152 for questions when dealing with anesthesia and surgery.

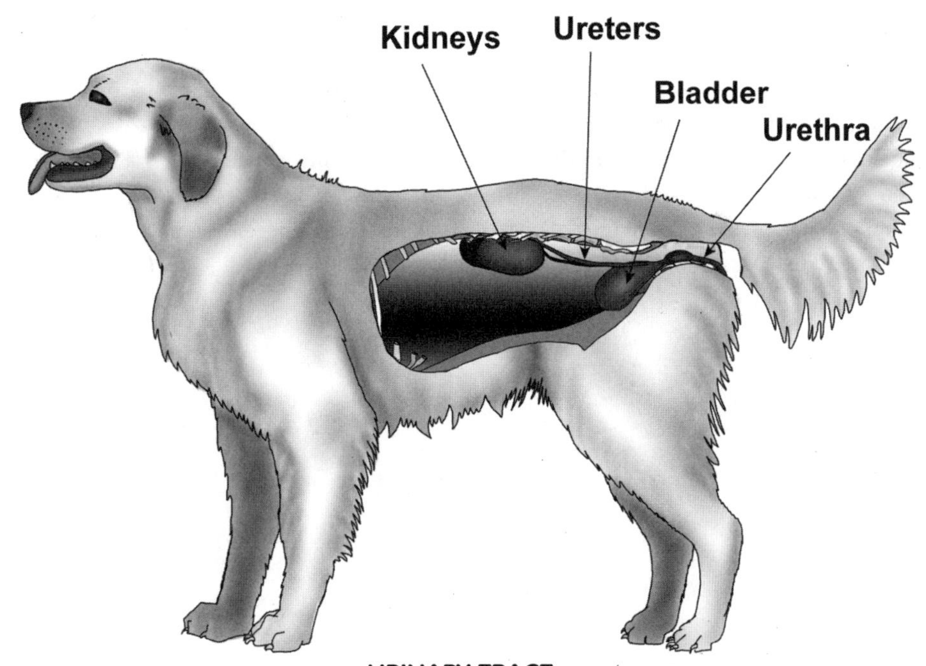

URINARY TRACT

Diagnosis: Urinary tract stones (urolithiasis, bladder stones, kidney stones, cystic calculi)

Urolithiasis refers to the formation of concretions (stones, uroliths, calculi) in the urinary tract. Forming in the kidneys or urinary bladder, they sometimes pass to lower sites (ureters, urethra). Stone formation results from urinary tract infection, metabolic disorders (excessive excretion of a substance in the urine), and dietary issues. Their composition depends on the underlying cause.

Urinary tract stones can cause inflammation and infection. Symptoms include: blood in the urine; frequent urination; and straining to urinate—although some dogs show no symptoms at all. Blockage of the urethra causes unproductive straining to urinate—be aware this is an emergency situation requiring immediate intervention (medical or surgical) to relieve the obstruction.

Abdominal ultrasound is the most useful technique for diagnosing kidney or bladder stones. X-rays are diagnostic in many cases, but some stones don't show up on X-rays. Contrast material in conjunction with X-rays or CT scans is sometimes used, especially when an obstruction is suspected.

Urinary tract stones are treated surgically or medically, depending on their location, size, and mineral composition. Long-term management is important to prevent recurrence. This includes: modifying the diet; prescribing medication; and encouraging increased fluid intake. Urine samples should be monitored frequently for acid/base balance, concentration, and the presence of infection.

Questions to Ask Your Vet

1. Where are the stones located?
2. If located in the urethra or ureter, are they causing an obstruction?
3. Does my dog have a urinary tract infection?
4. Do we know the mineral composition and/or the cause of the stones?
5. Can the stones be removed?
6. What can be done to prevent formation of new stones?
7. How frequently should my dog's urine be monitored?

Resources

Quick Reference: Monitoring Your Dog's Vital Parameters

Body temperature: normal range is 100.5 to 102.5 degrees Fahrenheit. Using a rectal thermometer (one made for human use is just fine), lubricate the end with petroleum or K-Y® jelly and gently insert approximately 1 inch into the rectum (it may be helpful to have a friend or relative hold your dog or distract him with an ear scratch). Keep the thermometer in place for about one minute before reading the temperature measurement. Note: ear thermometers, as used in people, don't provide accurate measurements in dogs.

Respiratory rate: normal rate for a dog at rest—not panting—is 18 to 34 breaths per minute. Count the number of breaths per fifteen seconds and multiply that number by four to determine your dog's breaths per minute.

Heart rate: the normal rate for a dog at rest is 70 to 140 beats per minute. Young puppies' heart rates are higher (up to 180 beats per minute), and in general, the smaller the dog, the more rapid the normal heart rate. Place your flat hand against your dog's chest wall just behind the armpit region (where a stethoscope is normally placed) to feel the heart beat. (This can sometimes be felt more easily when your dog is lying down on his side.) Count the number of beats per fifteen seconds and multiply that number by four to determine your dog's heart beats per minute.

Mucous membrane (gum) color: the normal color of the gums is pink. Some diseases can cause the color to become yellow (associated with jaundice), whitish (associated with anemia), or gray or blue (associated with some heart and respiratory abnormalities). Note: dogs can have pigmented gums, which prevents accurate assessment of mucous membrane color.

Capillary refill time: When momentary pressure is applied to your dog's gum, it will blanche (turn white), but then immediately—within one to two seconds—turn pink again. This process becomes prolonged (greater than two seconds) in dogs that are dehydrated, in shock, or have some types of heart disease.

Skin tenting: When the skin over the back of your dog's neck is tented (picked up between two fingers) and then released, it normally immediately returns to its original flattened position. When a dog is dehydrated, the skin remains tented.

Quick Reference: Toxic Substances

The list on the following pages contains some of the more common toxic substances and symptoms of their ingestion or exposure, but it *is not* all-inclusive. Immediately consult your veterinarian and/or the ASPCA National Animal Poison Control Center (888-426-4435, www.aspca.org/apcc) if your dog has ingested or had exposure to something unusual.

- **Do not** wait for symptoms to develop before making this call. If your dog requires treatment, the sooner he receives it the better.
- **Do not** induce vomiting at home before speaking with someone knowledgeable. Although it's tempting to rid the body of the toxic substance, in some cases, vomiting causes more harm than good, especially if the substance is caustic (irritating to the lining of the mouth, esophagus, and stomach).

Provide as much information as possible (i.e., description of the material ingested and the product container/packaging; amount ingested; your dog's approximate body weight, the number of dogs involved).

Plants	**Symptoms**
Amaryllis	vomiting, diarrhea, abdominal pain, depression, tremors
Azalea	vomiting, diarrhea, drooling, weakness, depression, coma
Castor bean	abdominal pain, drooling, vomiting, diarrhea, tremors, seizures
Chrysanthemum	drooling, vomiting, diarrhea, incoordination
Crocus	vomiting, diarrhea, collapse, shock, kidney failure
Daphne	vomiting, diarrhea, weakness, collapse, seizures, kidney failure
English ivy	vomiting, diarrhea, depression, tremors
Foxglove	vomiting, abdominal pain, heart rhythm abnormalities, collapse
Kalanchoe	vomiting, heart rhythm abnormalities
Marijuana	depression, incoordination, abnormal heart rate
Mistletoe	vomiting, diarrhea, incoordination, seizures, collapse
Mushrooms (wild)	vomiting, diarrhea, kidney failure, liver failure
Narcissus bulbs	vomiting, drooling, depression, seizures, heart abnormalities
Oleander	vomiting, heart abnormalities, collapse, coma
Peace lily	drooling, vomiting, difficulty swallowing
Rhododendron	vomiting, diarrhea, drooling, weakness, depression, coma
Sago palm	vomiting, diarrhea, depression, liver failure, seizures
Schefflera	drooling, vomiting, difficulty swallowing
Tulip bulbs	vomiting, drooling, depression, seizures, heart abnormalities
Yew	tremors, incoordination, labored breathing, heart failure

Human Foods	**Symptoms**
Avocado	vomiting, diarrhea
Chocolate	vomiting, diarrhea, hyperactivity, heart rhythm abnormalities, tremors, seizures
Garlic	red blood cell abnormalities
Grapes	kidney failure
Onions/onion powder	red blood cell abnormalities
Macadamia nuts	vomiting, depression, joint swelling, fever, weakness, tremors
Moldy foods	tremors, seizures
Raisins	kidney failure
Yeast dough	stomach distention, depression, disorientation, incoordination
Xylitol (sugar substitute)	vomiting, weakness, incoordination, seizures, liver failure, coma

Household Chemicals and Medications	**Symptoms**
Acetaminophen (Tylenol®)	vomiting, diarrhea, liver disease
Ant bait	vomiting
Anti-freeze (ethylene glycol)	kidney failure
Cigarettes	excitation, salivation, vomiting, diarrhea, collapse, coma, death
Cocoa mulch	vomiting, tremors, seizures, death
Compost	vomiting, tremors, seizures
Fertilizers	vomiting, diarrhea, liver disease (symptoms dependent on type of fertilizer)
Liquid potpourri	oral ulcers, drooling, vomiting, neurological symptoms (depression)
Mothballs	vomiting, anemia, liver disease, seizures, coma
Nonsteroidal anti-inflammatory medications (aspirin, ibuprofen, and other human products)	vomiting, diarrhea, depression, kidney failure
Pennies containing zinc (post-1982)	vomiting, depression, diarrhea, weakness, collapse, anemia
Rat/mouse poison	internal bleeding
Silica gel packets	vomiting, loss of appetite
Snail/slug bait (metaldehyde)	restlessness, vomiting, tremors, seizures
Toilet tank "drop-ins" (cleaning products)	vomiting

Recommended Reading: Books and Web Sites

ASPCA Complete Dog Care Manual by Bruce Fogle, DVM (Dorling Kindersley, 2006)
Complete Care for Your Aging Dog by Amy Shojai (NAL Trade, 2003)
Grieving the Death of a Pet by Betty Carmack (Augsburg Fortress, 2003)
Hound Health Handbook by Betsy Brevitz, DVM (Workman Publishing, 2004)
How Doctors Think by Jerome Groopman, MD (Houghton Mifflin Company, 2007)
The Merck/Merial Manual for Pet Health Home Edition, Cynthia M. Kahn, editor (Merck and Company, Inc, 2007)
PDR (Physicians' Desk Reference) for Nonprescription Drugs, Dietary Supplements, and Herbs (Thomson Healthcare Inc., 2007)
Pet Loss and Human Emotion: A Guide to Recovery by Cheri Barton Ross and Jane Baron-Sorenson (Brunner-Routledge, 2007)
Pets Living with Cancer: A Pet Owner's Resource by Robin Downing, DVM (AAHA Press, 2000)
Plumb's Veterinary Drug Handbook, 6th edition, by Donald C. Plumb (Wiley-Blackwell, 2008)
Veterinary Drug Handbook: Client Information Edition by Gigi Davidson and Donald C. Plumb (Wiley-Blackwell, 2003)

American Animal Hospital Association HealthyPet.com **www.Healthypet.com**
American Veterinary Medical Association Animal Health Brochures
www.avma.org/animal_health/brochures/dog_owners.asp
ASPCA Poison Control Center **www.aspca.org/apcc**
Cornell University College of Veterinary Medicine, Pet Loss Support Hotline
www.vet.cornell.edu/Org/PetLoss
Cornell University College of Veterinary Medicine, Free Animal Health Resources Web Sites
www.vet.cornell.edu/library/FreeResources.htm
National Center for Complementary and Alternative Medicine **www.nccam.nih.gov/health**

Advocacy Aids: Forms and Charts

Several different templates and forms that will help you excel as your dog's medical advocate are available via the *Speaking for Spot* Web site (www.speakingforspot.com). I invite you to download and use them to help you manage your dog's health-care needs. They include:

- **Emergency Contact Information Form**
- **Log of Current Health Issues**
- **Veterinary Hospital Visit Form**
- **Medication and Treatment Planner**
- **Contingency Plan Form** **(see p. 249)**
- **Log of Current Medications**
- **Health History Form**

Illustration Credits

Beth Preston (pp. 8–256)
Alexander Frederick (pp. 289–371)

Photo Credits

All photos in this book are by Sumner W. Fowler *except:*
pp. 24, 38, 106 by Blair O'Neil
p. 58 (the author and Boomer) and p. 130 (the author's daughter Susannah and Vinnie) by Alan Kay
p. 76 by Susannah Kay
p. 244 by Tonya Perme

About Sumner W. Fowler

Sumner has been Staff Photographer at the Marin Humane Society in Novato, California, since 1980. In addition to 30-years-worth of weekly portraits of adoptable dogs, cats, and other small animals, he has trained his keen photographic eye on many aspects of sheltering, including staff and volunteers, special events, children's summer camp programs, humane officer training exercises, special care and foster pets, and rescue operations. Sumner is an avid photographer of wild animals—especially birds—and farm animals, and also takes many commissioned pet portraits.

Sumner's award-winning work has appeared in 12 books, was featured in *Best Friends: Portraits of Rescued, Sheltered, and Adopted Animals*, and has graced numerous book covers. His images have been used by prominent humane organizations, published in numerous magazines, and reprinted by several greeting card companies.

Sumner has not only great technical mastery, but most importantly, empathy for his animal subjects. Not surprisingly, his favorite photographs "really illustrate the way I feel about the animals, and the way they feel about me. When I capture an animal's essence, I feel I've succeeded." Sumner's work has provided a visual chronicle of animals, and people working on their behalf, for almost three decades.

About the Marin Humane Society

Founded in 1907, the Marin Human Society (www.marinhumanesociety.org) is a progressive, award-winning animal shelter, offering refuge and rehabilitation to nearly 8,000 animals each year through myriad community services, including adoptions, foster care, behavior and training, humane education, lost-and-found pet services, low-cost clinics, and more.

Index

Pages in *italic* indicate illustrations.